Biohackers

BIOHACKERS

The Politics of Open Science

Alessandro Delfanti

PlutoPress
www.plutobooks.com

First published 2013 by Pluto Press
345 Archway Road, London N6 5AA

www.plutobooks.com

Distributed in the United States of America exclusively by
Palgrave Macmillan, a division of St. Martin's Press LLC,
175 Fifth Avenue, New York, NY 10010

British Library Cataloguing in Publication Data
A catalogue record for this book is available from the British Library

ISBN   978 0 7453 3281 9   Hardback
ISBN   978 0 7453 3280 2   Paperback
ISBN   978 1 8496 4906 3   PDF eBook
ISBN   978 1 8496 4908 7   Kindle eBook
ISBN   978 1 8496 4907 0   EPUB eBook

Library of Congress Cataloging in Publication Data applied for

This book is printed on paper suitable for recycling and made from fully managed
and sustained forest sources. Logging, pulping and manufacturing processes are
expected to conform to the environmental standards of the country of origin.

10   9   8   7   6   5   4   3   2   1

Typeset from disk by Stanford DTP Services, Northampton, England
Simultaneously printed digitally by CPI Antony Rowe, Chippenham, UK and
Edwards Bros in the United States of America

Contents

Preface

Every year a different Italian city hosts a large and important hackmeeting. In 2007 hackers were gathering in Pisa at the Rebeldia social centre: the centre's rooms were filled with weird computers, cables, hackers, political activists and free culture advocates. The hackmeeting is a place where people share knowledge, and workshops are organised in a completely open way with a wiki. That year, my friend Tibi and I held a workshop titled Hack Science, in which we intended to discuss with hackers the various ways in which social movements had got their hands dirty with science: used it, contested it, commissioned it, conducted it according to their political needs. That was perhaps the first time in which I explicitly linked science with hacking.

However, the description of the workshop said: 'when activism knocks on labs' doors'. Naïvely enough, our idea was not to talk about active hands-on intervention on the natural world, but rather the use of science for political purposes. Hackers were eager to discuss how to hack science, but despite the significant politicisation of the hackmeeting – workshops that year ranged from conspiracy to cryptography, from Linux to VoIP privacy – they wanted to discuss the possibility of getting their hands dirty with science. Then, and only then, science politics might follow, with its corollary of expert panels, independent research and protest.

Three years later, I was packing up my stuff and heading to the US West Coast to meet DIYbio (Do-It-Yourself Biology), a network of amateur biologists that in many ways is related to the traditions, myths and practices of hackers and who even share physical spaces with computer hackers (they set up wet labs for citizen biology in hacker spaces). Shortly before leaving Italy, I asked a friend to give me some classes on basic biotechnology tools and practices in order to refresh my knowledge from my days as a veterinary microbiologist and cytologist. The main piece of advice he gave me was about chemical and biological safety. He wanted to make sure I would not use dangerous chemicals such as methyl bromide (which he believed to be poorly regulated in the US) in an environment as creepy and unsafe as an amateur lab set up in a garage.

Yet when I visited hackerspaces in Seattle and Los Angeles I found myself extracting DNA from strawberries with a buffer solution made out of dish soap, or trying to use free software to resuscitate a ten-year-old polymerase chain reaction machine. I quickly realised that the science conducted by DIYbio might be very basic, and certainly not dangerous at all. Hacking in the sense Italian hackers gave to the word was only one part of their activities. On the other hand, I saw DIYbio members dealing with the FBI, organising conferences on open science, launching start-ups, looking for funding and writing letters to the US Presidential Commission on Bioethics. Politics for DIYbio were as routine as they are for any other social movement. Yet it was a very different type of politics compared to the radical tradition of Italian hackers.

These two stories illustrate how hacking can be a complex and multifaceted technical and political concept: that's why I believe that referring to hacking to explore biology helps make sense of some of the transformations that life sciences have gone through during the last decade. During and after the two events mentioned above, I have continued working on the edge between open science and forms of resistance to the new enclosures represented by intellectual property rights. As a crucial part of my political experience I knew that information and knowledge, far from being a common good freely created and shared by collective intelligence, were being increasingly privatised. Indeed, intellectual property rights have emerged as one of the main battlegrounds where the long-standing clash between privatisation and redistribution of wealth takes place. Scientific knowledge has been, together with the cultural industry, the main object around which this clash has revolved. Think about the problem of patents on genes. Yet, I also knew that the opposition between 'open' and 'closed' science was not enough to understand these clashes.

Profit, the organisation of labour and production, hierarchies and participation are problems just as important as access to information and knowledge. Thus, this book is not intended to be merely a sociological account of open science, but rather a proposal on the complex evolution of biology and its relationship with society and the market. Furthermore, I hope it will contribute to the debate about openness, free culture and hacking that has left the inner circle of practitioners long ago and has become diffused on a global scale.

I consider myself a member of the open access and free culture movements, and I am aware that my book runs the risk of being biased. Yet I also firmly believe a different viewpoint on open science has long been needed. Hacking, open source, piracy and free culture are all parts of the battles over information that are among the most important in contemporary societies. The radical request for transparency that characterises Wikileaks and governments' response to its practices, the rise of the Pirate Parties in Europe, the ghost of the hacker group Anonymous and its global actions, the incredibly harsh juridical clashes around intellectual property rights we have been witnessing over the last few years and the global regulations emerged to control them – all these phenomena testify to the growing importance of struggles around information control in our time.

Who controls the creation, distribution and appropriation of information and knowledge? This question is bound to become one of the key questions of our times and has deeply affected my work. Several years ago, while I was studying a transformation in the public image of scientists, I ended up (much to my surprise) tackling the relationship between cultural change and biocapitalism's evolution, where intellectual property rights and access to knowledge and information are crucial issues that have a deeper and more complex role than issues of public image.

But this is also a narration made up of three stories. In order to witness and report them, I undertook a journey whose stages included an animal health research facility in Padua, hackerspaces on the US West Coast, university rooms in Berkeley, pubs in Silicon Valley and finally dozens of webpages, online forums and listservs, as well as a long analysis of the Italian and international press and a study of two genetic databases. This journey has been curiously intertwined with my personal story. I began working on Craig Venter and his ship *Sorcerer II* while studying for my Masters in Science Communication, which was the first contact I had with research in science and society after a career in a completely different world. In fact, several years ago I briefly worked at an Istituto Zooprofilattico Sperimentale (IZS). Visiting the IZS delle Venezie, interviewing Ilaria Capua and writing Chapter 6 somehow reunited my current and my former professions. As a former colleague, I enjoyed meeting an Italian veterinarian who has acquired such an international dimension and yet keeps on working for an Italian public institution. Meeting the DIYbio crowd and hanging out with them was incredibly interesting and fun, and it allowed me to use some of my forgotten vet lab skills. As Mackenzie Cowell once noted, as an ex-biologist, open science advocate and hacking fan, I match perfectly the profile of your typical DIYbio nerd.

I am grateful to the groups and activists in the San Francisco Bay Area, Los Angeles and Seattle for sharing their curiosity and their time with me. Adam Arvidsson, Yurij Castelfranchi, Christopher Kelty and Nico Pitrelli had a special role in the genesis of this work and I cannot thank them enough for their advice, the intellectual challenges they posed me, the time they spent helping me, their support and their friendship. I would also like to thank Anita Bacigalupo, Michel Bauwens, Blicero, Beatrice Busi, Mariella Bussolati, Andrea and Mauro Capocci, Anna Casaglia, Gabriella Coleman, Magnus Eriksson, Andrea Fumagalli, Alessandro Gandini, Nikolaj Heltoft, Katie Hepworth, Marc Herbst, Steve Kurtz, Marina Levina, Paolo Ligutti, Lars Bo Løfgreen, Marco Mancuso, Federica Manzoli, Rachel Moe, Bertram Niessen, Helga Nowotny, Edoardo Puglisi, Roberta Sassatelli, Giulia Selmi, Johan Söderberg, Giuseppe Testa, Sara Tocchetti, Alan Toner, Penny Travlou, Fred Turner and many other friends and colleagues for helping me, sharing their ideas with me and commenting on earlier versions or parts of this work. David Castle at Pluto Press was an invaluable resource and believed in this project from the beginning, while two anonymous reviewers provided me with incredibly valuable suggestions and critiques. And of course this book would have never been written without Valentina Castellini's love, patience and support.

Parts of the material that now compose this study have been published during the last few years in *New Genetics and Society*, the *Journal of Science Communication*, the *International Review of Information Ethics*, *Studi Culturali* and the edited volume *Activist Media and Biopolitics*. The doctorate programme in science and society at the University of Milan and the Interdisciplinary Laboratory at SISSA (International School for Advanced Studies) in Trieste, gave me the space, the resources and the intellectual environment I needed to develop the ideas that compose this work. The Center for Society and Genetics at the University of California Los Angeles, where I spent six wonderful months, provided me with a relaxed and committed environment that was crucial for the materialisation of my PhD research. The *Journal of Science Communication* gave me the opportunity to work in an international project – something rare in Italy – and to contribute directly to the Open Access movement. Editing what is most likely one of the few open access journals in the field of science and technology studies was a crucial part of my professional growth. Finally, the Genomics Research and Policy Forum at the University of Edinburgh gifted me with a peaceful month away from my daily duties, allowing me to work on the first draft manuscript of this book.

I am also in debt to many people and experiences outside the academic world. The San Precario group and the EuroMayDay network are not only grassroots political projects on labour, precarity and welfare: they are also wonderful places where theoretical reflection and daily intervention walk hand-in-hand. I am grateful to the Italian hackmeeting community and the Los Angeles Bicycle Kitchen for showing me what an anti-authoritarian approach to technology means in practice, and to the Autistici/Inventati collective for providing me and thousands of others with free software tools and anonymous mailboxes, and for keeping online their more or less reliable services regardless of police searches, shutdowns or lack of money. The LASER group (Autonomous Laboratory on Science, Epistemology and Research) was for a few years an exceptional experience of contamination between different types of critical knowledge, and its work deeply influenced me.

Finally, I would like to remember Franco Carlini. He was a great journalist and public intellectual with an incredible political instinct and true commitment. Franco gave me a space in *Chip & Salsa*, the pages about technology of *Il Manifesto*, where I started working as a journalist and learned to keep a critical and political stance while writing on science, culture and technology. All these people and collective projects fuelled my interests in biology, intellectual property rights and information politics not only as intellectual challenges but, more importantly, as crucial political issues. I hope this work reflects how my view on science has been shaped by both the academic and political environments in which I belong.

Milano, Autumn 2012

1
Cracking Codes, Remixing Cultures

Yes, I am a criminal. My crime is that of curiosity. My crime is that of outsmarting you, something that you will never forgive me for.

'The hacker manifesto' (The Mentor 1986)

Crack the code, share your data, have fun, save the world, be independent, become famous and make a lot of money. There is a link between contemporary scientists devoted to open biology and the ethics and myths of one of the heroes of the computer revolution and of informational capitalism: the hacker.

In this book I show the existence of a confluence between the Mertonian ethos, the famous account of scientist's norms of behaviour proposed in the 1930s by the science sociologist Robert Merton (1973) and the hacker ethic, a very diverse and heterogeneous set of moral norms and cultural practices whose foundations are based upon the desire to have a free and direct approach to technology and information.

The hacker ethic emerged in the 1960s within the first hacker communities in the United States and while different versions of it have been formalised in several books, manifestos and writings, what makes hacking interesting today is exactly the wealth and diversity of practices and cultures it represents. The emerging open science culture I point out is influenced by this wealth, as it mixes rebellion and openness, antiestablishment critique and insistence on informational metaphors, and operates in a context of crisis and transformation where the relationship between researchers and scientific institutions, and their commercialisation and communication practices, are redefined.

In this book I refer to *biohackers* – life scientists whose practices exhibit a remix of cultures that update a more traditional science ethos with elements coming from hacking and free software. It is well-known that cultures related to hacking and free software are indebted to the modern scientific ethos. Yet what I want to show is how hacking and free software are now contaminating scientific cultures, in what we could somehow be defined as cultural feedback.

1

This process of coevolution is linked to the widespread and deep influence computers have on the scientific enterprise. In fact, the stories this book contains are related to the creation of genomics databases and community labs, and the use of online sharing tools and open source solutions.

Beyond the analysis of communication tools, I will explore a world in which the emergence of new scientific communities and new alliances between different actors are changing the landscape of scientific production. The sharing of genomic data through open access databases, the cracking of DNA codes, the standardisation of biological parts or the production of open source machinery for biomedical research represent one side of a process that also involves institutional change and challenges some of our assumptions about the relationship between research, commerce and power. A cultural shift lies at the centre of these transformations. Therefore, while one of the main problems analysed in this book is the widespread adoption of open access and open source solutions by biologists, my goal is to show that their relationship with hacker and free software cultures is deep and in some cases straightforward. In this way I tackle two main problems, one of which is the role of open science within the framework of informational and digital capitalism. The complexity of open science politics goes beyond the opposition between openness and closure and pushes us to look for a deeper understanding of today's transformations in biology. The other is the evolution of scientists' culture and how it interacts with the way science is done, distributed, shared and commercialised.

The three cases I present in this book are meant to exemplify the many different directions open biology is taking. Craig Venter, the US biologist known for his role in genetics' commercialisation and subjection to secrecy and intellectual property rights, sailed the world's oceans in order to collect genomics data and information he would then, for the first time, share publicly through open access databases and journals. The Italian virologist Ilaria Capua challenged the World Health Organization's policies on access to influenza data by refusing the institution's offers to upload its research group's data on avian influenza genomics on a password-protected database. Both Venter and Capua founded their own independent open access databases, although their goals were completely divergent. The rise of a do-it-yourself biology movement in the United States, DIYbio, was based not only on the American amateur science tradition, but also on explicit references to hacking and open source software from which it borrowed practices that it then applied to the life sciences.

I must stress that I do not use 'hacker' as a native category; in fact, most biologists that use open science tools and practices do not define themselves as hackers. Among the cases I present, only DIYbio has explicit relationships with the hacker tradition. In other cases, as will become clear in the following chapters, hacker cultures represent a source of innovation and contamination of scientists' cultures. Yet all three cases represent a move towards a more open informational environment and also a critique of the current system of the life sciences. Finally, they all have very different relationships to issues such as commercialisation, profit, or autonomy from institutions.

These cases are not meant to be interesting from the viewpoint of their scientific output – that would somehow go beyond my capacities. Also, it would be difficult to compare the scientific output of high-profile individual biologists such as Venter and Capua with such a diverse and decentralised movement as DIYbio, which may never become an important place for innovation. While I am aware that my choice could be seen as asymmetrical, I believe the juxtaposition of these three cases helps me to reach the main goal of this book. Indeed, by analysing both discursive strategies and socio-economic practices of contemporary biologists who use open science tools, I investigate their role in the changing relationship between science and society and try to give a multidimensional, stratified picture of the politics of open science.

The case studies I analyse are not impartial and are not generalisable, nor do they represent the whole spectrum of new open science practices, yet I argue that these biologists can all be a rich model for current transformations in both life sciences and informational capitalism. In particular, the culture to which I am referring gives scientists tools they can use in order to solve some of the political and societal problems raised by the increasing privatisation of genetic research by means of patents and other restrictions on accessing biological data. It can also be considered as an expression of a change in the institutional and socio-economic settings of contemporary biology: life sciences innovation now takes place in increasingly complex and mixed configurations, in which open data policies and open access tools coexist with different, and more strict, sets of access policies and intellectual property rights (IPR). Further, life sciences are now open to the participation of new actors, such as citizen scientists, start-ups and online collaborative platforms. These biologists have a role in hacking biology.

Hacking has an active approach to the shaping of the proprietary structure of scientific information – to who owns and disposes of biological data and knowledge. But it also poses a challenge to Big Bio[1] – the ensemble of big corporations, global universities and international and government agencies that compose the economic system of current life sciences – that aims at modifying the institutional environment in which biological research takes place by asking the question: who can perform biomedical research? Biohackers work against the high and well-defended thresholds to access that characterise Big Bio institutions. The skyrocketing costs of setting up a biomedical research laboratory, the increased complexity of biological research, the formal education required to work in a university or corporate lab, the complex bureaucracies that run scientific institutions, the legal and technological obstacles that prevent most people from accessing biomedical information have all been subject to attacks in the name of openness. Openness thus does not refer simply to access to information, but also to institutional change towards more open environments.

In fact, today the word 'open' has become an umbrella term that may refer to very different practices. In this book I use the expression 'open science' to describe a broad range of practices that include open source, open access, citizen science and online cooperative science or science 2.0. Intellectual property rights such as copyright, patents and trademarks grant owners exclusive rights to immaterial assets such as musical or literary works, inventions and designs. But during the last couple of decades alternative intellectual property rights have emerged as new forms of IP protection that allow widespread sharing and reuse. The term 'open source' refers to methodologies that promote the 'free redistribution and access to an end product's design and implementation details'[2] and in the biomedical sense should strictly refer to the use of legal licenses and technological platforms that allow access, sharing, reuse, recombination and modification of biomedical data such as genomic sequences. Yet 'open' has been used to define practices of free access and participation broadly (Hope 2008). Open access is an expression related to access to scholarly publications vis-a-vis the traditional system of journals that function according to a subscription fee model. There is an open science based on external collaborations or characterised by broader autonomy from institutions, for example, in the case of citizen science projects. Science 2.0 usually refers to any practice of cooperation carried out by means of online collaborative tools.

'Open science', in sum, includes all these different and somehow heterogeneous practices. Hence, when analysing open science I do not focus only on intellectual property rights, but more generally on the practices that foster access to the production of scientific information and knowledge. Thus, in this book we will embark on a journey through the very different ways in which science can become open and free. We will see how open science can be detached from the control of bureaucracies, but also represent a business and marketing model, and how it can widen citizens' access to scientific knowledge or even become a resistance practice.

THE TRAGEDY

The most common narrative about open science tells us that, once upon a time, science was an ethical enterprise: sharing, equality, disinterest and the common good drove the everyday work of scientists. Then evil corporations entered science and changed the rules of the game, patenting life, enclosing the commons, and eventually destroying the willingness to share data, information and knowledge. But today, so the story goes, we have new tools that together with the old open science spirit can be used to rebel against evil, defeat it and allow scientific knowledge to flow freely again. These tools are open source and open access science, and they can be used to tear down the barriers to the access of scientific knowledge.

The expression 'tragedy of the anticommons' comes from a paper published by *Science* in 1998 (Heller and Eisenberg). According to this formula, the proliferation of restrictions to access, patents and industrial secrets represents an obstacle to innovation. Michael Heller and Rebecca Eisenberg reverse the classic perspective on the 'tragedy of the commons', Garrett Hardin's 1968 work that has been used to justify the centralised management, or privatisation, of common goods. In a well-known passage Hardin stated that no pasture can be managed as a commons forever:

> Picture a pasture open to all. It is to be expected that each herdsman will try to keep as many cattle as possible on the commons. ... As a rational being, each herdsman seeks to maximise his gain. Explicitly or implicitly, more or less consciously, he ... concludes that the only sensible course for him to pursue is to add another animal to his herd. And another; and another ... But this is the conclusion reached by each and every rational herdsman sharing a commons. Therein is the tragedy. Each man is locked into a

system that compels him to increase his herd without limit – in a world that is limited. Ruin is the destination toward which all men rush, each pursuing his own best interest in a society that believes in the freedom of the commons. Freedom in a commons brings ruin to all. (Hardin 1968, p. 1244)

Hardin's position has been criticised from several perspectives. On the one hand, according to the definition by Nobel laureate Elinor Ostrom (1990), the commons should be interpreted as an 'institution for collective action' and only a hasty and individualistic approach can lead to ruin. On the other hand, the informational commons famously have distinctive characteristics. Information is said to be a non-rival good: no cattle can get through an informational pasture, for using a piece of information does not prevent anybody else from using it. In fact, intellectual property rights are artificial enclosures needed to extract value from a resource that is indefinitely replicable.

According to Heller and Eisenberg, and their diametrically different perspective on the commons problem, the increase of patenting in biotechnology inhibits innovation forcing actors to navigate a complex and atomised territory where intellectual property rights owned by several distinct parties raise the cost of doing research.[3] The cause of the anticommons effect is the fragmentation of property rights and the increased number and scope of barriers to access vis-a-vis the necessity of the 'assembling of an assortment of complementary bits of knowledge and research tools, each of which might be owned by distinct parties'.[4]

Furthermore, according to social studies of science, anticommons are also a symptom of the changes in the link between science, capital and society. During the last decades of the twentieth century the relationship between private corporations and academic science has become stronger, causing a general reconfiguration of the roles and dynamics of scientific research. Commodification is part of a major shift that has affected the social relations of knowledge production (Nowotny et al. 2001; Hedgecoe and Martin 2008).

Finally, the rise of anticommons has been interpreted as a cause of the corruption of the norms of good science, expressed by the adherence to corporate values and goals by the producers of scientific knowledge.[5] Patenting, secrecy and the quest for profit radically conflict with the norms of modern open science, namely with the 'commitment to the ethos of cooperative inquiry and to free sharing of knowledge' (David 2003, p. 3). The free and open dissemination of knowledge remains an important ideal associated with scientific progress. According to many authors and open access advocates, we need to couple the rise of new technological tools with a restoration of the modern open science culture. For Victoria Stodden, today's open science movement is not updating the social contract of science: 'what we're doing is returning to the scientific method which has been around for hundreds of years. It is what a scientist is supposed to do' (Stodden 2010b; see also Hope 2008).

The Budapest Open Access Initiative (2001), one of the main manifestos of the open access movement in scholarly publication, opens by combining the old open science culture and the new information and communication technologies:

> An old tradition and a new technology have converged to make possible an unprecedented public good. The old tradition is the willingness of scientists and scholars to publish the fruits of their research in scholarly journals without payment, for the sake of inquiry and knowledge. The new technology is the internet. The public good they make possible is the world-wide electronic distribution of the peer-reviewed journal literature and completely free and unrestricted access to it by all scientists, scholars, teachers, students, and other curious minds.

But my point is exactly that 'the old tradition' of the open science ethos is not enough to understand the transformations we are facing. This narrative about a corrupted Eden and its redemption is too simple and static. In order to present a different viewpoint on this story, I refer to the hacker ethic in order to study three open biology research projects. Criticising the main narration related to the cultural basis of open science, I want to shed light on the transformations that are shaking today's science: a new open science culture that is emerging among biologists, evolving from the twentieth-century Mertonian ethos but also comprised of several new cultural elements.

In his 1942 account of scientists' behaviour, 'The normative structure of science', Robert Merton famously proposed what is now a classic list of norms of behaviour which govern academic scientist's work and science's functioning. The norms that guide research practices, later summarised by the acronym CUDOS, are communalism, universalism, disinterestedness and organised scepticism. These imperatives are linked to rewards given to members of the scientific community who follow them, and sanctions applied to those who violate them. *Communalism* means that scientific data are a common good and need to be shared freely. Individual creativity must be recognised in the form of authorship, not ownership. *Universalism* means that science cannot use criteria such as race, religion or personal qualities to evaluate scientific claims. *Disinterestedness* is a norm against fraud and against the intrusion of personal interests in scientific activity. *Organised scepticism* states that the whole scientific community must be able to check facts and ideas until they are well-established and recognised.

Autonomy and disinterestedness are also two of the main characteristics of Michael Polanyi's 'Republic of Science' (1962). Polanyi uses the scientific community as a model for democratic societies. According to him, the free cooperation and self-coordination of scientists are directed towards the discovery of a hidden system of things, and the search for originality encourages dissent. The authority of the Republic is established 'between scientists', and not above them. But once established, authority does not need to be rejected. Rejection of authority happens during a crisis, during oppositional and controversial times in which scientists can decide to overthrow who reigns over the Republic.

Yet as historians and sociologists have pointed out, the Mertonian ethos is neither an accurate description of scientists' work nor a set of moral norms scientists should follow. More recent work in the sociology of science has identified a number of problems in Merton's proposal, namely in the importance of disagreement and controversies that are not deviations from a consensual norm but rather ubiquitous characteristics of the scientific enterprise (Collins and Pinch 1994; Laudan 1982). Furthermore, the norms of disinterestedness and objectivity can assume very different meanings for different scientists, and 'counternormal' behaviour that implies violations of Merton's norms are frequent and often rewarded (Laudan 1982; Mitroff 1974).

Thus CUDOS norms are rather to be considered a means for scientists to position themselves within a precise historical social contract between science and society, as they serve as an 'organizational myth of science' (Fuchs 1993). Several authors have tried to put Polanyi's and Merton's normative visions into a more sociological context, to both modernise and criticise them. The overall result is a significantly more complex scenario, in which autonomy and disinterestedness are not seen as values internalised by the scientific community but ways of positioning within a system of incentives that rewards them (David 2003; Krimsky 2006; Lam 2010; Ziman 2000).

Adrian Johns emphasises that the very idea of a pure science based on sharing and cooperative behaviours, that constitutes the basis of Merton's proposal, is one we owe to the debates about intellectual property, piracy and plagiarism and is related to the emergence of a corporate research sector that changed the image of proper science as a highly moral enterprise. Although premised on digital media, the open science movement's ideological foundations in fact 'date back to the mid-20th century patent conflicts and to the normative view of science that they generated' (2009a, p. 509). Thus the Mertonian ethos should not be considered as a prescriptive account of what a scientist should do (Merton himself was clear in that respect) but rather as a context-specific and adaptive cultural framework which provides scientists with strategies of action rewarded by a specific socio-economic configuration of scientific research.

Obviously, discursive strategies have always been of primary importance in the struggles that characterise the scientific field (Bourdieu 2004, p 77; see also Shapin 2008). On the other hand, hackers provide a multifaceted example of a range of cultures attuned to the economic dynamics of the software world made of start-ups, people escaping from academia, corporate networks, garages, online communities and computer science departments. As hacker cultures are an important component of contemporary informational capitalism, their role in changing open science might be crucial. The history of hacking is the history of the dream of a personalised technology enacted both through cultural and technical innovation.

But an attempt at clearly separating the many facets of hacking from each other might prove to be flawed, as it seems impossible to demarcate the line of separation between hacking as a critical and alternative approach to corporate computing, as a community-driven technological evolution and as a source of socio-technical innovations to be subsumed by corporations and governments. The countercultural roots of hacking and its contribution to the IT industry are inextricable.[6] Thus hacking is useful in developing an understanding of the similarities and differences between the approach to scientific institutions, corporations, intellectual property rights and antiestablishment critiques expressed by the biologists I have included in this study. Pointing out the relationship between scientists and hacking also allows me to make a comparison between open science on the one hand, and the history and the political economy of free and open source software on the other.

OPEN SCIENCE POLITICS

The stakes of open science are high and recognised widely. Information and knowledge circulation have always been critical battlefields not only for science and technology but for human societies more generally. Problems related to the access of information and intellectual property rights have often been the cause of controversies and battles in the history of communication technologies and information societies. The results of those battles were foundational for the evolution of today's capitalism (Johns 2009a; Mattelart 2003). Yet the rise of digital media and new information and communication technologies have magnified and generalised the importance of information control and management: information is a crucial commodity in the global market and has a key social and economic role.

In scientific research, data access has long been recognised as a central issue in the very definition of the purposes and nature of the scientific enterprise (Hilgartner and Brandt-Rauf 1994; see also Nowotny et al. 2001). Among the main indicators of this shift are the introduction of massive amounts of private capital into scientific research and the extraordinary growth of intellectual property rights that has taken place over the last few decades. On the one hand, we have witnessed the broadening of intellectual property rights beyond their classic reach, exemplified by the patenting of genetic sequences. On the other hand, we have seen the rise of a global intellectual property regime modelled after the United States' legal framework and designed to protect private and public capital invested in biomedical research and development.

The blurring of distinctions between university and corporation research is one of the effects of changes mandated by the World Trade Organization (WTO) in intellectual property regimes, cuts in public expenditures for research and increased funding for targeted research by multinational corporations (Mirowski and Sent 2008). These changes have triggered battles over information control and property that have been particularly harsh in the life sciences field. A couple examples are: the developing 'world war' against bioprospecting, and the legal and political clashes around gene patenting that have been taking place in several countries as well as on a global scale. But this wave of science commercialisation and increase in intellectual property rights has also caused a response, namely the creation of the open science movement which is part of a broader free culture movement. However, the crisis science is going through is not confined to problems of access to information and knowledge, but includes battles over the very shape of science's institutions as well as over public participation in and control of scientific knowledge production.

The changes I depict in this book show how these conflicts, and the emergence of new open science practices, are making science unstable again, and how unpredictable the outcomes of such battles are. In conflicts that revolve around information, what is at stake is the very balance of power that characterises the digital economy, which is becoming increasingly contested and open-ended. Hence, new social actors and practices related to information control have acquired a peculiar prominence, and no analysis of today's capitalism can avoid facing the challenges they pose, as they once again inflame remaining tensions between public and private, and between autonomy and centralisation. Of course, crises always involve opportunities and allow a glimpse at profound transformations.

In this book, then, I analyse scientists' role in these battles related to openness, control and commercialisation, as well as in the emergence of new interests and economic models, and I ponder what future directions open biology will give to the evolution of information societies. As Adrian Johns underlines in his book on the history of piracy, not the mere analysis of technological or juridical change, but rather the study of the dynamical interlacing of technologies and society, can underpin a revision of the relation between creativity, commerce and power (2009a, p. 517).

In fact, there is no need for another book about the feasibility and desirability of the open science turn. Janet Hope (2008), Michael Nielsen (2012) and Peter Suber (2012), just to name some of the most recent and recognised works on the topic, have written persuasive books on the need for a more open approach to science production. Broadly speaking, open science advocates argue that more openness and transparency would make science more productive, more inclusive, more democratic, or might eliminate it from the influence of selfish private interests.

Yet a critical understanding of new open science practices needs to avoid the approaches that inform most of the accounts of these transformations. Indeed, I describe the scientific enterprise in a form radically different from both hegemonic accounts of open science and open access, which are often flat and spoiled by uncritical enthusiasm, and scholarly work on academic capitalism (Mirowski and Sent 2008). In my view, life sciences fully embody the diversity of free and open source software politics, which range from radical critiques against science commercialisation to new forms of neoliberal openness. By criticising and putting into context hegemonic visions of open science, I highlight that the opposition between enclosure and openness is inadequate for describing today's controversies related to power over information and knowledge.

Furthermore, these conflicts are shaping the meaning of concepts that lie at the very core of information societies' development, such as creativity, openness, access and property. Analysing the changes that involve the control and management of scientific information and knowledge in the age of digital media is then an important step towards understanding the transformations our societies are undertaking.

In this sense, the life sciences are particularly interesting. On the one hand, biomedicine is intended to be the scientific field in which openness is the most valuable and culturally entrenched. The moral stakes of making biomedical knowledge open and accessible are easily recognisable. On the other hand, biology is a privileged object of inquiry in studying the relationship between science and society. Studying biomedical sciences allows us to dig deeper into the evolution of capitalism itself.

The analysis of biosciences' role in driving the evolution of capitalism could probably be traced back to Michel Foucault. As Melinda Cooper maintains, the biotech revolution needs to be understood in the larger context of the evolution of the neoliberal project, in which it had a special role. They both share an ambition to overcome limits to growth through 'a speculative reinvention of the future' (2008, p. 11), and life's annexation within processes of accumulation was part of a recreation of capitalism that took place in the 1990s and was based on the expansion of intellectual property rights. Furthermore, the history of hacking, computers and software is integrated fully into the history of neoliberalism and into the development of informational capitalism. At the same time, battles around information technologies and intellectual property are part of a tendency towards the democratisation of information societies.

Open biology is embracing values and practices taken from the world of hacking and free software. This means that science is experiencing the same type of differentiation and complexity shown by hacker cultures. In fact, what is interesting about hacking is precisely its political heterogeneity and diversity. The hacker ethic is not a formalised set of moral norms. Levy (2010a) lists elements such as: access to computers should be unlimited and complete, all information should be free; mistrust authority; hackers should be judged by their hacking, not bogus criteria such as degrees, age, race, or position; you can create art and beauty on a computer; computers can change your life for the better.

Yet while some authors, such as Pekka Himanen (Himanen and Torvalds 2010) provide a rather crystallised picture of the hacker ethic, in the last few years a group of scholars has highlighted the enormous degree of plurality that makes hacking cultures so interesting. Of course, hackers are portrayed in ways that provide several different moral evaluations of their ethics and practices: hackers can be portrayed as young kids with pathological addictions to the Internet, informatic criminals, tinkerers, or geeks committed to information freedom. Ethnographies of hacker communities have shown how hacker cultures are contended, continuously negotiated and reformulated, and how we should stay away from a moral binary that tends to classify hacking as a unidimensional moral activity to be either lauded or blamed.

According to Coleman and Golub, hacker morality 'exists as multiple, overlapping genres that converge with broader prevailing political and cultural processes' (2008, p. 256). This diverse constellation of genres revolves around the idea that the very action of coding is a moral act. Hence, in societies in which politics and moral codes are significantly linked to digital technologies, hacking can be a privileged site of observation. Christopher Kelty asks us to resist the urge to classify hackers under a constituted political category, since hackers 'are involved in the creation of new things that change the meaning of our constituted political categories' (2008, p. 94). Similar experiences, such as hacklabs and hackerspaces, can reveal different, or even opposite, political genealogies (Maxigas 2012).[7]

Studies of hacking cultures thus enable a greater understanding of the codes and evolutions of liberalism itself, as hackers question basic assumptions about power over knowledge and information. Yet while they highlight the constitutive ambivalence and complexity of hacker cultures, those studies often lack a more general approach to the evolution of information societies. The contamination of science with hacking and free software cultures can provide useful and fresh insights into this evolution, as applying the complexity and heterogeneity of hacker politics to biology allows us to highlight how openness, far from being a unidimensional characteristic of modern science, should be seen as a multidimensional practice related to contemporary neoliberalism.

Negotiations around concepts such as openness, autonomy or commercialisation that take place in the life sciences can be an important means through which to transform information societies and digital capitalism. During the twentieth century, open science's values were associated with those of liberal democracies – Robert Merton pictured science as a 'blueprint for democracy', while Michael Polanyi identified science as the model for a 'free society' (Merton 1973; Polanyi 1962; see also Thorpe 2008). Yet today's life sciences are increasingly associated with free market values (Cooper 2008; Sunder Rajan 2006).

Sharing of genomics data, biomedical knowledge and scientific information is moralised in new ways and the very meaning of terms such as 'openness' and 'freedom' are continuously negotiated. In his work on piracy and its relationship with the emergence of the modern concepts of authorship and intellectual property rights, Adrian Johns highlights creative work as a heavily moralised enterprise. The ethos of openness and access 'is upheld as virtuous because [it is] true to the intrinsic character of genuine science' (2009b, p. 56).

Open biology is ambivalent. While open data sharing and rebellion against bureaucracies are crucial elements of a critique of the current regimes of science – characterised by increased privatisation, commodification and unjust power distribution – it would be naïve to see open science as a pure liberatory programme. Far from being only a tool of resistance against science commercialisation, open biology is participating in the evolution of neoliberal societies. Open science, though, is not merely a case of capitalism's recuperation and instrumentalisation of a critical culture. Rather, I suggest it is one of the elements that drive the very evolution of a *new spirit of capitalism* based on openness, sharing, autonomy and horizontality (Boltanski and Chiappello 2005). Thus the normative approach to open biology that is prevalent among open science advocates might obscure power relations and new forms of accumulation.

As a member of the free culture movement, I think that adopting the kind of critical approach that has now become common in free and open source software studies – where the many facets of 'open' information and knowledge production have been analysed in depth – would be useful for the evolution of open science itself. A former hacker and now arch-enemy of the hacker and free software movement, Bill Gates, stated in an interview that if he were a teenager today he would be hacking biology (Levy 2010b). Certainly hackers represent a challenge to current incumbents. Yet this does not mean open biology is good per se, but rather that we should problematise the complex and diverse sets of practices that characterise it.

For example, open science practices can be part of an effort to establish new business models not based on patents, but on the ability to manage open data circulation. Therefore, we need to focus on several different political levels not limited to the opposition between openness and closure. After all, the history of information societies is soaked with neoliberal claims about the 'free flow of information' as well as autonomy and freedom from governmental constraints (Harvey 2005). In liberal societies, different forms of openness related to information and knowledge management had a role in the definition of new territories of accumulation. Adrian Johns, for example, highlights how the clash between British pirate radio stations and the BBC provided a new moral basis for liberalism. Laissez-faire, commercialisation, distributed creativity and freedom of inquiry somehow became 'orthodoxies' in the 1980s and 1990s (Johns 2009b, p. 58). More than a century before, British anti-patent campaigns would state that if the national industry were not to decline, free trade principles had to be extended to the realm of creativity (Johns 2009a, p. 272). In his history of neoliberalism, David Harvey portrays an ideal state of access to information under free market conditions, in which there are no asymmetries that interfere with individuals' rational economic decisions (2005, p. 68).

REMIXING CULTURES

Current biologists' cultural material is at first represented by the set of values inscribed in the famous Mertonian norms, the classic twentieth-century scientist's ethos. These values represent a toolkit or a repertoire scientists can use to build strategies of action. According to Bourdieu, 'researchers' strategies are oriented by the objective constraints and possibilities implied in their respective position' (2004, p. 35). These strategies also depend on the structure of the field they operate in and on the distinctive dispositions of their habitus.

Nevertheless, the strategies they pursue are not intended as conscious plans leaning towards pursuing specific interests, but rather as 'persistent ways of ordering action through time' (Swidler, 1986, p. 273). A set of cultural norms supplies 'culturally-shaped' skills that render us 'active, sometimes skilled users of culture' (ibid, p. 277). Therefore, cultural frameworks both enable and constrain individual choices and actions. They are limiting but flexible, stable but not static toolkits that actors can reconfigure and into which they can insert new tools coming from different cultures.

But they are also resistant toolkits, that are somewhat rigid and do not allow users to modify and twist them to their liking. Individuals have access 'to only a small, not unlimited number of alternative regimes of action and justification coexisting in a state of instability' (Silber 2003, p. 430) from which a person can choose a specific configuration of cultural resources. Individuals can reappropriate cultural competences that originated in a given historical context, modify them so they are adapted to new circumstances and use them in new strategies of action. Cultural norms can survive the phenomena that generated them, as in the case of the Mertonian ethos surviving the socio-economic configuration of twentieth-century science that enabled it.

But they must be updated by inserting elements coming from new and different cultural traditions. Like new music created by remixing elements coming from two original songs, this cultural mash-up produces new harmonies and rhythms, and opens as many opportunities as it closes. The functioning dynamics of science described by Robert Merton or by Michael Polanyi are still widespread and vital within today's public representations of science, even though the world which sustained them has radically changed. The boundaries between public and private (for profit and non-profit research increasingly blur), and corporate values burst into academic science even if the continuity with the old ethos does not fail. According to Alice Lam, an author who has studied the ethos of contemporary scientists who work on the 'fuzzy' boundaries between academy and industry, one should not predict a shift in the work practices of scientists towards an entrepreneurial mode without taking into account the fact that it can happen only 'within a strong continuity of the "old" academic frame as actors mix disparate logics at the blurred boundaries between institutional sectors' (Lam 2010, p. 335).

As Steven Shapin (2008) puts it, 'people matter'. The personal virtues and the ethos of contemporary scientists are central to understanding their practices and institutional relationships, and they are not merely a matter of public perception unrelated to the material development of science. Thus, scientists matter, since 'the history of science is invariably told through the lives of its heroes' or its rebels – and this book is no exception (Harman and Dietrich 2008, p. 1).

Furthermore, in the R&D and entrepreneurial networks where the forces that drive science development and capitalist economy gather – where technoscientific futures are made – the role of scientists' personal virtues and personalities reaches its zenith. Scientists push science forward and give direction to technoscientific development, and their culture assumes a special role in contexts of crisis and change – like that characterised in the clash between different intellectual property rights approaches, the privatisation of genetic information and the emerging open science movement.

Ann Swidler suggests that 'unsettled lives' are moments in which the reprocessing of an already existing culture enables new strategies of action. In a contested arena, cultural models are more explicit and they can directly shape action, influencing chances of success and therefore the opportunities different actors are able to seize. Luc Boltanski and Laurent Thévenot use the expression 'critical moments', in which people 'realize that something is going wrong ... that something has to change' (1999, p. 359).

The disputes that originate during these breaking points are subject to an 'imperative to justify', namely the need to justify one's actions and to display the reasons behind one's criticisms in a legitimate and generalisable way. It is during those critical and unstable moments that a scientist can mobilise an ethical system as a justificatory regime and reconfigure it by inserting new tools coming from different cultures. Thus, in this work, I analyse the role of this emerging and remixed ethos that provides biologists with new tools and strategies of action that can be used to overcome some of the challenges they face during their participation in today's scientific enterprise.

In Chapter 2 I outline the history of open science and of the involvement of biology in a wave of legal, political and societal clashes around the rise of intellectual property rights. I then, on the one hand, link this phenomenon to the shifts we have witnessed during the last 40 years in the way scientific research is organised, funded and circulated: the socio-economic configuration of scientific research has become complex and multifaceted, with the blurring of the boundaries between academic and corporate science and the coexistence of several intellectual property rights approaches. On the other hand, I contextualise open science within the emergence of different innovation modalities, such as open online production, peer-to-peer, social production and the theories that tackle the clash between the rise of informational common goods and the eternal attempts at private appropriation.

The goal of Chapter 2 is to include science in the processes through which the circulation, property and management of information and knowledge shape society and capitalism. In this sense, the relationship I propose between genetics (and genomics) and information and communication technologies is not accidental: biotechnology genesis partially overlaps computer and hacking history. For example, they share common birthplaces (MIT, San Francisco Bay Area), while the so-called post-genomic biology, subsequent to the announcement of the sequencing of the human genome in 2000, is heavily dependent on hardware (computational power, databases) and software (programmes to analyse and extract relevant information from genetic sequences). Information and communication technologies had an important role in the development of genome mapping and sequencing technologies, as well as in more recent subdisciplines such as synthetic biology, epigenetics, or metagenomics, and in commercial enterprises such as the direct-to-consumer genetic testing industry. The informational metaphors that surround genetics have shaped scientists' approach to genetics and have had a role in the rise of intellectual property rights in this field (Kay 2000; Keller 2000; Waldby 2000).

Switching to the role of scientists' culture and ethos, one of the underlying questions of my work is: who/what is a scientist today? The scientist as a public figure who acts outside institutional structures of scientific publishing (such as scientific journals, conferences and workshops) to address the mass media's general public is a recurrent figure of science history. Furthermore, many scholars have emphasised that the current configuration of science is characterised by more complex negotiations and conflicts between consumers, social movements, enterprises and academic research.

This reconfiguration mirrors more general transformations that have been depicted with labels such as post-industrial society, post-Fordism, knowledge society, informational capitalism or reflexive modernisation (on those distinctions see Kumar 2004). However, such diagnoses share some elements, such as the centrality of the media and communication.

This has deep implications for the public communication of science (Bucchi 1998; Nowotny et al. 2001). Cultural changes are not merely staged in private spaces. In fact, the media provides the main arena in which scientists can show their personalities, moral values and ways of participating in the building of a future that is embedded in the scientific enterprise. Rae Goodell (1977) used the expression 'visible scientists' to describe researchers with a high media profile, able to make the first pages of the newspapers and manage their relationship with the press better than other colleagues. These scientists are irreplaceable characters for the contemporary public, since their role is to introduce science directly into the main public arena of industrialised societies: mass media.

Studies on the public communication of science and technology have underlined that several images of scientists coexist and circulate in popular culture. These alternative views of scientists can be diverse, opposite, heterogeneous and complex. A scientist can be a genius, a good guy, even a national hero, a dangerous mad scientist, a Victorian gentleman, a bureaucrat, a political activist, a rebel, a rockstar, a nerd, a villain, a maker, a citizen.

In Chapter 3 I introduce the figure of the hacker and propose a comparison with the modern scientist in order to build the basis for the analysis I conduct in the following chapters. In particular, I begin by presenting a history of hackers and an analysis of the development, role and importance of the hacker ethic within the more general framework of the so-called 'new spirits of capitalism'. These represent an update of Weber's thesis on the cultural and religious foundations of capitalism (2003 (1905)) and tackle the cultural basis of informational, digital capitalism that embodies ideologies of liberation, horizontality, sharing, cooperation and participation. I then link the cultural traits of the hacker to an analysis of two other characters: the rebel scientist and the profiteer scientist. In doing so, I hope to be able to shed some light not only on the evolution of science or of free culture, but also on the importance of hacking for the study of the evolution of information societies.

My argument is supported by empirical research based on three case studies distributed across the United States and Italy and located in the post-genomic era – that is, the decade that began after the announcement and the publication of the sequencing of the human genome in 2000. In the 2000s, in fact, the need for new forms of profit extraction pushed the bioindustry to include sharing as a form of information management and control. Furthermore, technological and political changes paved the way for the diffusion of open source and open access practices in the life sciences. The selected cases represent highly mediatised research projects that received attention by the press and produced a huge amount of communication material destined for the general public.

Second, the problem of access to and sharing of the data emerged as a crucial public issue in the communication production related to these cases, and they represent, in different forms, innovations and changes in the relationship between 'open' and 'closed' biology. Open science tools such as open databases, open access journals and open platforms for data sharing are used.

Third, these biologists operate in different and often opposing institutional settings: one is a scientist working for international public health institutions and with no relation to the life sciences private sector; another is a geneticist known for being the emblem of science's privatisation, of the corporate invasion of the life sciences and of the new enclosures on scientific information and knowledge such as patents and secrecy; the last is a community of amateur biologists external to the boundaries of science's formal institutions.

The first case involves the *Sorcerer II*, the Craig Venter Institute's research ship that circumnavigated the planet between 2003 and 2007 to collect, sequence and classify marine microbial genomes. The results of the Global Ocean Sampling Expedition were published in open access journals, and the data collected were deposited in a open access database called CAMERA. For the first time the 'bad boy' of science Craig Venter, famous for embodying a new type of scientist–entrepreneur, used open science tools. For this research project he put together several different types of scientific actors, from Google to the National Institutes of Health (NIH), and from the *Discovery Channel* to multiple universities (Delfanti et al. 2009).

The second case includes Ilaria Capua, an Italian veterinarian virologist who works within Italy's national public health system. In 2006, during the global avian flu crisis, she engaged in a public clash with the World Health Organization (WHO) over its policies on restricted access to data. A letter to her colleagues started a debate involving both scientific journals and the general press. Two years later, Capua founded an independent open access database under the umbrella of GISAID (the Global Initiative on Sharing Avian Flu Data), and the WHO eventually changed its policies. As a result, Capua became a famous open access advocate and her public image switched to one of a 'rebel' and 'revolutionary' scientist.

DIYbio (Do-It-Yourself Biology) is a network of amateur biologists established in 2008. It is now composed of several groups in major US (and European) cities. Their aim is to provide citizen scientists with cheap and open source tools for biological research which is to be conducted outside the boundaries of scientific institutions. In 2010 they started several collaborations with local hacker spaces to set up small labs. DIYbio also launched BioCurious (a biohacker space to be opened in the San Francisco Bay Area) and the OpenPCR project to build a polymerase chain reaction machine under open source principles. In these two years DIYbio has established dialogues with universities, companies, media and the US government.

To examine these cases I collected communicative materials from national and international media, scientific journals and press offices, and analysed data from multiple sources such as journalistic articles, TV interviews, scientific papers, press releases and websites. Then, by means of theoretical and qualitative discourse analysis, I focused on the images of scientists and his/her norms, virtues or ethics. In the case of DIYbio, the media analysis was coupled with a four-month participatory observation on the US West Coast and interviews with several prominent members. I also conducted an in-depth interview with Ilaria Capua and visited her laboratory in Padua. Finally, in addition to media analysis and interviews I analysed the socio-economic ecologies of these same cases: their economic alliances and scientific collaborations.

A NEW OPEN SCIENCE

The case studies included in this book have been chosen to show the complexity of the new open science phenomenon. Craig Venter is a corporate-oriented biologist famous for his unscrupulous use of patents and secrecy, for example during the race against the Human Genome Project. His shift to open science is very interesting because it shows how openness can be part of a corporate (and marketing) strategy. Ilaria Capua shows how public rebellion based on a reassertion of the modern science ethos pushed one of the last Big Bio institutions to shift to open access policies. Here, power over access to data, and not money, is at stake. DIYbio is an example of the direct transfer of hacker ethics in the realm of the life sciences, and it keeps discourses of participation and political critique together with the search for new business models.

These three cases share a common but flexible culture based on openness, but which include hostility against bureaucracies, extreme informational metaphors that to refer to DNA, and so on. Yet they show that this culture can be reconfigured in very different ways in order to adapt it to different needs, such as those faced by a freelance private scientist working for profit, by a 'public' scientist struggling towards shifting the balance of power over data, and by a very complex movement such as DIYbio that includes discourses of participation and autonomy, disruptive business models and so on.

The individual elements I found are not new. Yet their remix is innovative and creates a new and emerging figure of the scientist, one who uses open science tools more attuned to the current configuration of the relationship between science and society, but who also rebels against bureaucracy and claims independence from academic and corporate institutions. Autonomy, independence and openness coexist with other elements – for example: a radical refusal of interference coming from Big Bio incumbents; the belief that bare information is good per se, as long as it is shared and accessible; the importance of being an underdog; an intense relationship with the media; the rebellion against the mechanisms of scholarly publishing and peer review; in some cases an explicit drive towards profit and entrepreneurship.

I suggest that these case studies represent a remix between an old culture that is pre-existing, accepted, embodied in a complete set of practices and norms and ready to be used, and a more recent ethos linked to several other fields of innovation. The justifications they produce guarantee scientists a fun and fascinating job, while at the same time working for the common good. The strategies they pursue are often related to the norms of behaviour attributed to the hacker. In this sense, individuals can mobilise ethics when the need for a reconfiguration of different cultures becomes more pressing. Contemporary scientists can still use some cultural elements belonging to the old Mertonian science ethos, since the influence of that culture has survived the social dimension from which it was born, but they often need to remix it with new and different ethical and cultural elements.

The complexity of open science politics lies in the spaces of possibilities opened by this confluence. Using the hacker ethic as an analytical tool has allowed me to highlight some of the elements shared by very diverse types of scientists: one of the biologists I studied belongs to public research institutions; the second is a free rider who drains money from venture capitalists, media companies and public agencies; and the third group are amateurs external to official science but immersed in a complex entre-preneurial environment. At the same time, the hacker ethic has allowed me to indicate important differences in their approach to information sharing, corporate models and institutional settings. The public dimensions of these biologists are related to the current configuration of the relationship between science and society, enterprise, universities and other actors which participate in the making and marketing of contemporary biology.

This emergent class of biohackers is related to a new type of interaction between scientists' practices and biology's social contract. I want to stress here that I am not referring to their ability to provide more productive models or to produce better scientific knowledge. What I would like to show is that the new open science social contract they prefigure and contribute to building could restore some of the sharing practices that characterised twentieth-century academic research. However, it would also be transformed, broadened and improved by web technologies and the widespread diffusion of open and peer production. At the same time, it would include more strict intellectual property rights regimes. Different forms of information management and control would coexist in an environment inhabited by creatures as diverse as companies, universities, public agencies, start-ups and new institutions such as citizen science projects.

The new open science culture linked to this social contract maintains a political ambivalence. Through their mobilisation of ethics scientists better position themselves within the current socio-economic configuration of biological sciences. Both academic and industrial research (provided that it is still possible to separate them clearly) have increasingly been using diverse and mixed approaches to intellectual property, and in some cases – such as database management – strictly proprietary models are seen as no longer sustainable.

Thanks to the open and free input of non-experts and voluntary contributors, the participatory processes of governance and the universal availability of the output, open and peer production might prove to be more productive than centralised alternatives. Thus open biology is not only a tool wielded against the current status quo and against the enclosures represented by secrecy and strict intellectual property rights. The way in which information circulates has important political consequences, and the role of new media as a tool for democracy is an important discourse underlying the whole development of information societies.

For example, photocopy machines (or online social media) can be seen as a metaphor for an open society when used by the illegal political opposition in an authoritarian country, which happened in Hungary in the 1980s (Dányi 2006). On the other hand, in a world in which openness, flexibility and freedom from bureaucracies and cooperation are elements that belong to a capitalistic mode of organising labour and production, we must rethink any easy commitment to open science as good per se and face its complexity.

Thus, biohacking can be an intervention in the marketplace as well as a practice of resistance. The case studies I present are to be considered as part of a shift towards a more open environment for biological research – open meaning both 'open to more participation and cooperation' and 'open to a more diverse set of modes of capitalist appropriation'.

2
Forbidden, Public, Enclosed, Open Science

Picture a pasture open to all.
    Garrett Hardin, 1968

Open science and its historical and economic basis are the main subject of this chapter. Focused on the political economy of the *production of scientific culture*, it will serve as an introduction to Chapter 3 in which I will tackle the *cultures of producers*. After outlining a short history of the emergence and establishment of open science and its crisis due to the new enclosure movement in the late twentieth century, I link it with the software realm and the rise of the free software and open source movement. In order to understand the genealogies of open science, I briefly analyse openness and participation in relation to their role as resistance practices and as parts of a new and emergent form of appropriation by digital capital.

The invention and institution of new ways of including more and more things in the realm of property are important parts of capitalism's development. Intellectual property can be seen as a way of subjugating immaterial entities such as inventions to new property regimes. For example, the reification of 'genetic capital' emerged from transformations of the very notion of what an animal is, such as the rise of property rights for cattle breeders in eighteenth-century Britain. While the notion of a 'gene' did not exist at the time, the rising awareness created by books and pedigrees that certified breeding animals as unique individuals with a particular commercial value represented a shift that somehow preceded the application of property rights to genes and genetic sequences (Brewer and Staves 1996).

During the nineteenth century in the United States a similar effort was put into obtaining legal recognition for plant breeders' rights (Kevles 2011). Yet the emergence of a new proprietary regime of science – what has been referred to as the 'second enclosures movement' (Benkler 2006) – is linked to the transformations of contemporary capitalism, where information and knowledge assume a leading role in profit accumulation. Marx defined primitive accumulation as 'the historical process of divorcing the producer from the means of production', (1990, p. 875) a process which preceded and made possible the specific mode of production of capital. His examples are the expropriation of the English rural population, the enslavement of American natives and the like.

Updates to the Marxist theory have proposed that primitive accumulation appears every time capital needs to find new ways out of a crisis in its mode of appropriation (Hardt and Negri 2000). Thus today's new, but still 'primitive' accumulation that Marx would probably call the 'original sin' of informational capitalism is characterised by enclosures that do not block access to the informational pasture but rather increasingly manage, adjust and control the flux of data and knowledge. Producers are not being divorced from the means of production, which are more and more diffused and used for free cooperation processes – millions of personal computers connected to the Internet – but the fruit of that cooperation is certainly being expropriated. Yet the periodic recreation of capitalism is accompanied by the imposition of the new limits inscribed in a new property form: in the case of biocapitalism, the exploitation of a new 'surplus life' (Cooper 2008).

The waves of propertisation that have characterised science arose from different types of property and different modes of appropriation. In this chapter I will retrace the history of open science that I outlined in Chapter 1, highlighting the links between different socio-economic settings that accompanied some phases of modern science and the practices scientists were adopting during those same phases. After abandoning the tradition of secrecy that characterised it during the Middle Ages, science has gone through the openness permitted by patronage (either private or governmental) in modern science, the enclosures expressed by the rise of intellectual property rights in the last few decades of the twentieth century, and finally the counter-attack of new open science movements in the early twenty-first century coupled with new modes of appropriation based on giving, openness and sharing. These phases, or tides, of openness and closure characterise the entire history of science as a form of communicative action, a history related to the different ways of appropriating and valorising information and knowledge.

OPEN SCIENCE TIDES

Open science is a method for producing scientific knowledge by spreading its results and opening them up to the revision of the entire scientific community, maximising information and knowledge circulation and sharing. At the opposite extreme, there is a 'closed science', a secret one or one in which communication dynamics are limited within the walls of an institution or subject to the payment of a license such as a patent or a copyright. However, the concept of science as a pursuit of public knowledge, which today may seem obvious, is actually the result of complex and stratified social and economic dynamics.

In their historical accounts of the rise of modern open science, William Eamon (1990) and Paul David (2001) have emphasised the social dimension of openness. The practices that compose open science, such as disclosure of knowledge, methods and data, peer review and cooperation are described as fruits of a social process rather than individual choices and predispositions. Open science is also a relatively recent phenomenon that has surfaced in the modern age. In the Middle Ages, science was secret knowledge not subject to public disclosure. An ethos of secrecy encouraged scholars not to reveal their knowledge. The 'secrets of nature' were not to be revealed to anybody but the select few who could pursue the penetration of those secrets.

Obviously, that ethos of secrecy was linked to the political problem of maintaining the social order – for example, the need to contain thinkers who wanted to question the link between religious knowledge and state power. But social and economic factors were also important, such as the role of guilds of engineers and craftsmen which opposed the open communication of technical secrets. Only in the seventeenth and eighteenth centuries, as a part of the scientific revolution, did open science emerge as a way of considering science as public knowledge, and as such significant questions were asked about the limits to freedom of inquiry.

David (2003) explicitly defines it as the 'open science revolution'. Early modern ideals pushed towards the open circulation of new ideas, and the ethos of secrecy was slowly replaced by a new ethos of making scientific knowledge public and open to scrutiny – a process also driven by science patronage and its search for public recognition. But the tradition of forbidden and secret knowledge also clashed against a more productive and fast-accumulating way of searching for the key to the secrets of nature. The Baconian ideal of scientific progress as a progressive and collaborative effort concurred with the shaping of public science: to Bacon, not individual genius but an egalitarian scientific community was at the centre of the scientific enterprise.

At the end of the seventeenth century, the newly founded Royal Society helped to institutionalise public science by fostering cooperation among scientists and by establishing a system of open communication composed, at first, of a record book and a correspondence system: the free dissemination of knowledge took centre stage. Scientists were not alchemists devoted to secrecy anymore, but gentlemen who pursued knowledge as a journey through Nature's wonders and in embryonic, non-bureaucratic institutions. Chris Kelty (2010; see also Dyson 2009) describes Victorian gentlemen with words such as magic, paternalism, wisdom and eccentricity: men (and only men) who worked in a circle of friends, for pleasure and pure knowledge, for exploration and peers' delight. For Steven Shapin, who gives the example of Charles Darwin as a gentleman scientist, the integration of science into structures of power and profit was never more than partial in the nineteenth century. The figure of the man of science as an amateur, conducting inquiry without expectation of a remunerated career, did *not* disappear (2008, pp. 41–2).

Later, in Europe and then in North America, the emergent role of the modern state in directing scientific activities was a crucial factor in the establishment of science as an enterprise based on public knowledge coupled with the professionalisation and institutionalisation of scientists – the transition from science as a calling to science as a job. The state was eager to mobilise more resources and apply scientific inquiry on a scale that altered the scope and, most of all, the character of scientific practice. Technological factors were at play in this transformation: the advent of printing was one of the main factors behind the scientific revolution and the shift towards the establishment of open science. The technological innovation represented by print brought with it new economic incentives linked to the emergence of new publics for science. Thus the emergence of a market for printed books made it obvious that not only reputation and fame, but profit as well, could be had by publishing one's secret knowledge in order to share it with a growing readership (Eamon 1990; Johns 2009a).

Yet until World War II funding for research came either from corporations or charitable foundations, in a modern form of patronage that shaped science's relation with society. Edgar

Zilsel (1945) has linked the emergence of the modern concept of scientific progress to the rise of capitalism: the ideal of science as a cumulative, collective and open inquiry was useful in establishing modern innovation systems. In that respect, the boundaries between industry and the state quickly became sharper. In fact, in the second half of the twentieth century, the relationship between open and closed science took a well-known form, strictly tied to the needs of industry and the state.

While in Chapter 3 I will further problematise this unidimensional account of science's functioning, here I would like to stress how in modern times and in university-based research communities open science was linked to a rather explicit pact between science and the state, one that allowed scientists to autonomously pursue basic inquiry in exchange for the disclosure of their findings to the system of technological innovation represented by private corporations. Two different organisational regimes, coexistent and complementary, were in place, and although they overlapped in many regards they were roughly based on two different regimes of knowledge sharing. On the one hand, a sphere where science was supported by public funding and private foundations' patronage, and thus was open, accessible and shared. On the other hand, a sphere of scientific research organised and funded by private entities for commercial profit and under proprietary regimes (see David 2003, p. 2).

The modern relationship between the open Republic of Science and the proprietary Realm of Technology was ratified by Vannevar Bush's report on the adaptation of the state's role in driving big military scientific projects during the war towards a new form of state patronage of science. A new social contract between state, research and industry guaranteed the stability of the kind of open science we continue to take for granted today. Freedom of inquiry, together with institutional autonomy and scientists' disinterestedness, were important ingredients of Bush's recipe for scientific progress and, of course, US national prosperity and security after the end of World War II:

> We must remove the rigid controls which we have had to impose, and recover freedom of inquiry and that healthy competitive scientific spirit so necessary for expansion of the frontiers of scientific knowledge. (Bush 1945)

Scientific progress on a broad front results from the free play of free intellects, working on subjects of their own choice, in the manner dictated by their curiosity for exploration of the unknown. Freedom of inquiry must be preserved under any plan for governmental support of science.

On the other hand, Robert Merton's norms of communalism, universalism, disinterestedness and organised scepticism portrayed and favoured the role of academic scientists as producers of public knowledge, open to public circulation and public scrutiny (also see Chapters 1 and 3 for a critique of this model). Several authors criticised this sharp separation between academic and industrial science, depicting a scientific system where the practices of these two actors were mixed, often similar and always complementary.

Furthermore, in more recent times scientific knowledge production has been subject to a radical reorganisation linked to the increasingly blurred distinction between private and public science, the rise of private funding and the more porous border between industrial and academic science. In opposition to a 'normal' form of organising discipline-based scientific research within academic settings and with public funding, these new forms of knowledge production are context-driven, based on multidisciplinary collectives and focused on problem solving.

Scientific research driven by the needs and money of industry embraced proprietary practices exemplified by the increased use of patents (see Gibbons et al. 1994; Ziman 2000). Even if, over the last century at least, patent rights have been strengthened and have embraced more kinds of subjects, since the late 1970s we have witnessed growing and broadening efforts to enforce the use of intellectual property rights to legally protect scientific knowledge. The rise of university patenting that had begun in the 1930s and accelerated in the 1970s reached its culminating point with two events that occurred in 1980 and are usually cited as emblematic of this 'appropriation shift' emanating from the US and quickly involving most Western countries (Johns 2009a; Popp Berman 2008).

One event was the passage of the Bayh–Dole Act in the US, followed by similar laws in an international context. It allowed the filing of patent applications for the findings of publicly funded research projects, and gave rise to the explosion of intellectual property offices in universities. The other was the Diamond v. Chakrabarty sentence, which extended patent laws to living material – namely genetically modified bacteria (US Supreme Court 1980).

Industry had an active and important role in this shift. In fact, there has been a continuous expansion of the role of industry in advocating stronger and broader intellectual property rights in the life sciences (Dutfield 2003). Industry's lobbying activity had a leading role in both the Diamond v. Chakrabarty trial and the passage of the TRIPS agreement (the Agreement on Trade-Related Aspects of Intellectual Property Rights) in 1986. Furthermore, corporations struggled to strengthen IPRs on living matter and have seen convergence in industries such as chemicals, seeds, pharmaceuticals and biotechnology.

As for genomics, one of the first major features of the growing commercialisation of this emerging scientific field came in 1991 when Craig Venter, a researcher at the US National Institutes of Health (NIH), announced he had filed patents on thousands of partial complementary DNA (cDNA) sequences associated with coding genes. Following the criticism that emerged because of these patent applications, Venter left NIH in 1992 and founded his own institute with private capital: The Institute for Genomic Research (TIGR). In the period from 1996 to 2001, when the first draft of the human genome was completed, again including a prominent role from Craig Venter (see Chapter 4), the rate of patent applications increased rapidly. The 'genome gold rush' had begun, as many firms entered 'a race to make claims on potentially valuable genes before the full sequence was placed in the public domain' (Martin et al. 2010, p. 151).

Such a way of organising knowledge production has also been described in terms of changes in the values that drive academic research efforts, that become more contaminated by industry goals. As one of the most important accounts of this shift puts it:

> universities can adopt 'values' from the corporate culture of industry, bringing forth an entirely new type of academic entrepreneur. Conversely, big firms adopt some of the norms of academic culture, for example when they give employees sabbaticals or provide other forms of training possibilities. (Gibbons et al. 1994, p. 37)

For John Ziman (2000), the CUDOS values proposed by Robert Merton were challenged by the emergence of a specular set of norms he summarises in the acronym PLACE: knowledge production becomes *proprietary*, *local* (it meets technical needs), *authoritarian*, *commissioned* and is based on the role of *experts* as problem solvers. Thus, according to this vision, science needs to create an active relationship with the different economic, political and social actors that influence and drive its development. The autonomy and economic security that allowed modern science to prosper are not guaranteed unless science is able to gather consensus within society. Finally, a more general turn towards flexible, just-in-time, informatised production was linked to this transformation. A paradigm for research and development related to information technologies was emerging, one that 'increasingly replaces one dominated by the technologies and organisations of mass production and consumption' (Gibbons et al. 1994, p. 125; see also Mattelart 2003 and Kumar 2004). 'Big science' was replaced by flexible and diverse collectives which included universities, start-ups, foundations, private companies, patients' associations and so forth, and in which intellectual property rights assumed an economic and organisational centrality.

After 2001 though, in a descending curve of that appropriation wave, the rate of patent applications decreased significantly, mostly because of the decline in university patenting (Leydesdorff and Meyer 2010). A general decline in the commercial value of this kind of intellectual property is not enough to explain this phenomenon. The business model that was in place before the human genome sequencing, namely the selling of access to databases, often proved to be unsustainable. Firms switched instead to drug discovery and development, or to selling services linked to the search within the huge amount of data generated by massive genome sequencing projects (Martin et al. 2010). This process occurred in parallel to a social backlash suffered by the main actors of the appropriation shift. Since the 1990s, a wave of social, political and legal clashes hit biotechnology. The accusation of betrayal of the social contract between science and society echoed in social movements against 'patents on life' and 'biopiracy' and was an important part, for example, of the global social movements which arose after the protests held in Seattle in 1999 against the WTO (an example of an emblematic text belonging to that movement is Shiva 1999).

At the same time, new forms of open science have emerged and now coexist with academic capitalism and corporate science (Gruppo Laser 2005; Hope 2008; Nielsen 2012). And once again, this latest wave of open science practices is linked to the coming of a new medium – the Internet – and to new social configurations. Information technologies have extended the possibilities of producing, sharing and using scientific data and knowledge in open ways. Science, like many other human activities, has experienced the great consequences of the technological revolution based on the Internet. New information technologies indeed contribute to the changing of the geographic frontiers of research – among scientific disciplines and between scientists and other citizens. The past few years have seen the explosion of scientific data publication forms, exploiting new IT technologies which have been made available to everyone – in a quick, convenient and free way – as the result of research projects. Scientific journals and open access archives are indispensable to online collaborative science, and the data they contain are the raw material that so-called 'science 2.0' is based on.

The movement for open access in scholarly publishing has produced an explosion of open access journals that have challenged the business model of traditional scientific journals, which are based on copyright and expensive subscriptions. There is a positive tendency towards open access publishing: according to a recent study, about 20 per cent of scholarly literature published in 2009 is freely available online (Björk et al. 2010). One of the main examples is the non-profit publishing group Public Library of Science (PLoS), which publishes several online life sciences journals freely accessible by anybody. With its PLoS One, Public Library of Science is also experimenting with a form of 'open peer review' for scientific papers, which allows the participation of the entire community of researchers: scientists can comment, correct and discuss the work of their colleagues, giving birth to a process of continuous revision of the published articles that has now been adopted by other publishers. ArXiv is a huge online, open access archive maintained by Cornell University and cofunded by the US National Science Foundation where scientists, especially physicists, mathematicians and computer scientists as well as bioinformatics, upload the preprints of their scientific papers. ArXiv has become the standard for publication in many disciplines, to the point that in 2002–2003 the Russian mathematician Grigori Perelman (winner of a Fields Medal), not being interested in pursuing an academic career, has used it to publish the papers in which he solved the century-long puzzle of the Poincaré conjecture, thus reaching the global mathematics community while refusing offers to publish with prestigious journals.

New online platforms are not the only innovation being used for setting up open science initiatives. As in commons-based peer production models, technological, legal and cultural changes are at play. For example, crucial to the spread of science 2.0 are legal methods to promote collaborative research. Science Commons, a branch of Creative Commons (CC), is an example of an effort to apply the 'copyleft' model (a play on the word copyright which refers to alternative licenses that grant the right to distribute copies and modified versions of a work) to scientific data and knowledge by creating licenses that allow users to access, copy, modify and redistribute scientific works or data without paying any royalties to a copyright or patent owner. According to Creative Commons, 10 per cent of the world's entire output of scholarly journals is CC licensed. Public agencies have adopted broad open access policies.[1] The NIH's public access policy states that:

> the public has access to the published results of NIH funded research. It requires scientists to submit final peer-reviewed journal manuscripts that arise from NIH funds to the digital archive PubMed Central upon acceptance for publication. To help advance science and improve human health, the Policy requires that these papers are accessible to the public on PubMed Central no later than 12 months after publication.[2]

Other major national and international agencies and institutions have slowly shifted to open access policies. Telethon Foundation, the Italian organisation which funds research on muscular dystrophy and other genetic diseases, has adopted mandatory open access policies for the research groups it funds.[3] Others comply with the Human Genome Project Bermuda Principles, that mandate not only open access to articles and research reports but free availability of genomic sequences data in the public domain.[4] Furthermore, in the life science corporate world we have witnessed the spread of innovation models in which open sharing of data is gradually becoming a form of information management that is complementary to stricter practices such as secrecy and the use of intellectual property rights (Bauwens et al. 2012).

This change is also brought about by the shifts that have occurred in the realm of genomics: the largest databases in the world are now open access, and private enterprises sell services linked to the management of raw data. 'Open' and 'closed' models of data management coexist both in the private and the public sectors. In 2010, for example, the British pharmaceutical company GlaxoSmithKline made a database publicly available that contained the structures and pharmacological data for 13,500 molecules that might possibly become new drugs against malaria (Anonymous

2010). According to Janet Hope (2008), open source business models are spreading in the biology sector, and a full-grown open source biotechnology is likely to rise thanks to the convergence of public sector, non-profit entities and private companies that rely on open science practices. Over the last few years, for-profit open science companies have sprung up in the United States. New companies such as DNAnexus or NextBio make their genomics data publicly available and often collaborate with, or are expressions of, bigger biomedical corporations. Besides the role of sharing practices as ways of managing information, some authors' roles are suggesting they might also represent a form of new corporate activism (Hope 2008; Sunder Rajan 2006) as well as a way of accumulating financial capital and media exposure.[5]

PEER-TO-PEER SCIENCE

Changes in contemporary science – which are closely linked to the innovations introduced by the use of the Internet – are complicating the picture. The very definition of 'open science' is at stake because of the emergence and spread of cooperative web platforms. Science is increasingly conducted outside the boundaries of scientific institutions, often in ways that might remind us of Peter Kropotkin's descriptions of workers' inventive activity (1996).

As a communication enterprise, science 2.0 practices go beyond information and knowledge sharing within scientific communities. Indeed, science is increasingly being produced and discussed by way of online cooperative tools by web users and without the institutionalised presence of scientists. Citizens conduct, discuss and circulate research outside the so-called ivory tower of science. Their radical claims for openness and access to scientific knowledge are heating up a debate on the boundaries of contemporary science. On one side, citizen participation means that science's decision-making procedures are at stake; on the other, scientific enterprise itself is changing. These worlds are increasingly engaging in an inter-communication, and the frontier between open and closed science should now be reconsidered: can web instruments really generate the collaborative non-hierarchical processes among peers that Yochai Benkler (2006) defines as 'commons-based peer production'? The web, an indispensable resource for contemporary science, is not only a technological instrument, but also a field in which different views collide on what science is and what its social purposes are.

Thanks to the Internet, citizen science is becoming more diffused. But it is not just a matter of diffusion: web tools are creating and facilitating new ways for lay people to interact with scientists or to cooperate with each other. Several definitions have been used to describe this phenomenon, such as 'citizen science' or 'do-it-yourself science' (DIY) – indeed, we are not entering a well-established world, but rather an emergent phenomenon still looking for stability. Here, I will use the definition 'peer-to-peer' (P2P) science (Delfanti 2010).

According to Michel Bauwens, P2P science is an attempt at restoring and broadening the lost openness of the scientific enterprise, as it allows citizens to get together and contribute to the production of scientific knowledge thanks to processes that start 'from the free contributory individual, not from a group-based negotiation of interests' (2010, p. 2). But what happens when lay citizens or non-scientists go online and engage in scientific activities?

'Popular science' or 'citizen science' are two traditional ways of defining grassroots science produced outside the walls of laboratories. The history of citizen science can be traced back to the very beginning of scientific knowledge production. See, for example, *A people's history of science* by Clifford Conner (2005), which is a long account of lower-class innovation from prehistory to computer hackers. Social studies of science and technology include an entire wave of scholarship on user-led innovation, or lay and popular knowledge.

However, the Internet has changed the way of collecting, sharing and organising the knowledge produced by people – peers – who do not belong to the established scientific community. Obviously it is not just a technical matter. If these emerging practices are still immature and difficult to grasp it is because they are the fruit of the recent convergence of several technical, cultural and social phenomena. The first example is the emergence of a technical and legal infrastructure which enables free online cooperation. The Internet is characterised by horizontal and pervasive diffusion, open protocols and collaborative tools enhanced by open licenses such as CC or common pools of knowledge available as resources freely available online. Second, we are witnessing the spread of practices, such as 3D printing, that show how online social production can move 'from bits to atoms' and be embedded in material goods.

Today we are witnessing the use of open source models in domains far from software – such as hardware, design, politics, money and so on. A famous and successful example in the hardware domain is Arduino, the Italian 'open-source electronics prototyping platform based on flexible, easy-to-use hardware and software'.[6] Regarding the political layer, there is the diffusion of a request for participation in science's dynamics which dates back to the 1960s that is still growing (see Jasanoff 2003). This is an important topic for science communication, and it was acknowledged by means of a shift towards more participative, multidirectional and inclusive communication practices. This convergence has resulted in an increase in the number of people who can produce or discuss scientific knowledge without any formal recognition as scientists, and in the way the Internet enables collaborative systems for them to interact and participate in these activities.

User-led and peer-to-peer science include very diverse ways of engaging in scientific knowledge production. The first type of P2P science is online discussion. It can be done via web tools such as blogs, independent forums comprised of patients, activists or amateur scientists and social networking websites. These spaces can be hybrid forums where citizens talk with scientists, or P2P spaces where non-experts have free discussions, exchange information and produce knowledge. Other examples are open online encyclopaedias such as Wikipedia, where anybody can contribute to a scientific entry without needing any formal qualifications, or open textbooks and notebooks where lay people can contribute to the stabilising of knowledge.

The second area of P2P interaction with science is represented by data collection, processing and analysis for a centralised institution. This includes the sharing of personal data – for example, in websites such as Google Health or social networks for data sharing such as those implemented by personal genomics companies such as 23andMe or other providers of medical and health services. In other cases, 'netizens' are asked to give some of their computers' computational time to process data within distributed computing projects. Examples are initiatives such as Folding@home or Rosetta@home, which rely on thousands of individual contributors who download software that uses part of the computational power of their personal computers in order to analyse and predict protein structure. These distributed computing projects can be as powerful as big mainframe supercomputers.

Other types are based on a request for distributed and active participation of the analysis of data that are collected and processed in a centralised way. For example, projects such as Galaxy Zoo, where volunteers are asked to classify galaxies by checking thousands of pictures taken by telescopes. Professional teams don't have the time and resources to analyse the increasing amount of information produced by these research projects, while small contributions by thousands of amateur volunteers have proven as effective as expert classification once aggregated through software tools. Galaxy Zoo has produced several scientific papers in high-ranking journals.

Finally, centralised projects ask a distributed network of citizens to collect independent and original data to help researchers. This is the case of the BioWeatherMap initiative, rooted in a broad network of volunteers that are asked to help build a distributed environmental sensing effort by sending microbial samples coupled with weather data. BioWeatherMap aims at understanding the geographic and temporal distribution patterns of microbial life. This third area of user-led science is composed of completely independent and community-driven P2P science projects which design research, perform experiments and analyse data with the support of distributed networks and platforms. The hobbyist scientists network DIYbio, which attempts to hack biology and promote garage biotechnology, is the subject of Chapter 6 and one of the most important examples of citizen science. MyDaughtersDna is an open platform for the sharing of information about genetic pathologies with researchers, physicians and patients; the Pink Army Cooperative is a non-profit co-op 'operating by open source principles' which works on personalised medicine for cancer: 'the first DIY pharmaceutical company'.[7]

THE FREE SOFTWARE LEGACY

Besides claims regarding the role, scope and results of open science with regard to values such as democracy or participation, openness in science is part of a broader movement towards open, horizontal, peer-to-peer production models, both online and offline. In fact it is part of a broader shift that began with the rise of free software in the 1980s and evolved with the emergence of open source software and the spreading of these models, outside the domain of software, to other fields of information and knowledge production. However, I'm not interested in analysing the technical and legal differences between free and open source software, a much-debated issue in the software world. This distinction has important political consequences for the way communities that produce different types of free, libre, open source software (FLOSS) perceive and define their practices. I will instead focus on a few features related to the political economy of these two modes of production, namely their relationship with property, private appropriation and economic justice.

Open source refers to a form of property which is not organised around the right to exclude others from a good or a service but rather around the right to use, share, distribute and modify an informational good such as a piece of software code (the source code). This free source code, being open, public and non proprietary, supports the existence of a community of users who are also its developers and sharers. The best known example of FLOSS is of course the operating system Linux, a set of various projects maintained by a community of users and developers thanks to a copyright agreement that guarantees free access to the source code of software such as Debian, Apache or Ubuntu. The famous meaning of 'free' – as in free speech, not free beer – is the basis of the organisational system of free software projects: openness to participation in the community, data transparency (the source code), freedom to share, use and modify code.

In order to analyse the link between this model and the scientific enterprise, I want to underline that the FLOSS model is not confined to the software domain. The rise of effective and large-scale cooperative efforts based on peer production of information and knowledge is typified by the emergence of free and open source software. At the same time this model is expanding way beyond software production, into practically every domain of information and cultural production. Among the typical examples one can list encyclopaedias, entertainment, news and textbooks.

Several authors and scientists have drawn an explicit (and often normative) comparison between the two worlds of free software and open science: free and open source practices and open science *must* converge in order to overcome the anticommons effects brought about by patents, industrial secrecy and copyright, and enhance the cooperative spirit of science (see Benkler 2006; Hope 2008; Stodden 2010b; Willinsky 2005). The modalities of innovation that emerged within the software field would then spread into other sectors of research, innovation and knowledge production. According to Benkler, peer production models at work in the scientific world are a great example of convergence between the free software methods and a different field of innovation, and he offers examples such as PLoS and arXiv in order to explain how this class of commons-based, non-proprietary production solutions to problems is giving birth to a whole area of information production and exchange unhampered by intellectual property (2006, p. 313).

Victoria Stodden supports this position, urging science to adopt a much needed open source approach in which 'open code is an important part of this, as much as open data' (2010b). Stodden's proposal is related to reproducibility: a publishing standard which includes analytical tools, raw data and experimental protocols, giving any scientist the possibility of reproducing a colleague's experiment. The Stanford biologist Drew Endy, a famous actor in the field of synthetic biology and open science, states that 'in 15 to 30 years something really interesting will develop between these two poles: FLOSS and synthetic biology'.[8] According to Endy, both companies and individuals will be able to make key innovations outside the walls of universities and corporations thanks to the diffusion of FLOSS models within biology.

The most important study on the link between open source and biology is *Biobazaar*, a book dedicated to 'the open source revolution in biotechnology' that applies the peer production model to life sciences (Hope 2008). Hope explicitly draws a comparison to FLOSS practices and the life sciences, arguing that 'none of the differences between software and biotechnology constitutes an insurmountable obstacle to implementing an open source "biobazaar"' (p. 189). For other scholars, the obvious link between free software and science can be exploited the other way around: Mertonian norms of communalism, universalism, disinterestedness and organised scepticism should contaminate the software field. Indeed, free software is also said to be a milieu where Mertonian norms become 'goals in practice' (Kelty 2008 p. 271) and where scientists, engineers and geeks are reinventing norms of behaviour. In particular, free software (and copyleft projects such as Creative Commons) is a place where these norms are embodied in technical and legal practices.

Money and reputation, both in free software and in the scientific field, would be two different but related currencies that participate together in the incentive system that sustains software production. Yet this does not prevent free and open source solutions from being part of capitalist appropriation models. While the commons-based peer production described by Yochai Benkler is a non-proprietary and non-market system of innovation and sharing, open source has positive results for grassroots, distributed and non-profit projects which are part of a transformation towards more liberal and egalitarian societies, as well as for private corporations seeking profit.

In a speech at the World Social Forum in Porto Alegre, Manuel Castells (2005) maintained that open source is an 'a-capitalist' practice that can be adopted by resistant, autonomous communities as well as by private corporations driven by profit. Thus open source is not necessarily anticapitalist: a claim that might seem obvious but that contradicts many enthusiastic approaches to commons-based software production as a site of the subversion of capitalist relations of production. It is very easy to prove that many companies, including very large corporations, practice open source as one of the several possible different approaches to intellectual property. According to Castells, open source is 'a-capitalist', meaning that it is compatible with very different social logics and values. Yet open source surely represents a new form of information production which, being based on a peculiar social organisation that does not rely on the incentive of profit nor on exclusive rights to use a good, has profound political implications. Open source 'may affect the way we think about the need to preserve capitalist institutions and hierarchies of production to manage the requirements of a complex world'.

Janet Hope, in her work that is explicitly dedicated to the open source model in biotechnology, goes further, arguing that software freedom 'is a means to an end. That end is free competition'. Open source licenses are 'essentially procompetitive: they promote low barriers to entry and dismantle the monopoly powers associated with intellectual property rights' (2008, p. 167). Indeed, the similarities between the FLOSS movement's discourses and capitalist discourses of flexibility, free market and information flow, have been underlined by several authors who maintain that 'the open software movement cannot be entirely dismissed as a significant alternative to digital commodification and techno-capitalism' (Best 2003a, p. 466).

In their analysis of the relationship between different political uses of the FLOSS philosophy, Gabriella Coleman and Benjamin Mako Hill (2004) argue that the political ambiguity of FLOSS goes beyond the lexical ambiguity of the words free or open, and that 'the interplay between FLOSS philosophy and practices as it travels through multiple social, economic and political terrains may reveal more than (first) meets the eye'. For example, they underline how words such as freedom and openness are appropriated in different contexts by different users of FLOSS practices, such as non-profit software projects (Debian), technology corporations (IBM) and anti-corporate activists (the Independent Media Centers or Indymedia).

Free software communities, indeed, are usually not interested in overthrowing capitalist rules but rather in writing good software and not keeping it private or secret. Still, free software can be a nightmare for corporations that use strictly proprietary models. In October 1998 internal memos leaked from a Microsoft executive, which later became known as 'the Halloween documents', were diffused through the Net by the open source advocate Eric Raymond, making known the most powerful software corporation's fear of the Linux operating system. According to the document, free software was a major threat to Microsoft's position in the operating system market thanks to its ability to mobilise developers and users:

> The ability of the OSS process to collect and harness the collective IQ of thousands of individuals across the Internet is simply amazing. More importantly, OSS evangelization scales with the size of the Internet much faster than our own evangelization efforts appear to scale.[9]

GOOGLE YOUR GENES

Openness, giving and participation. In contemporary advanced capitalism, opposite and complex forces drive the role and evolution of openness and participation practices. In order to highlight the main forces that are relevant to this study I shall go back to 1922, when Marcel Mauss published his most famous work: *The gift: The form and reason for exchange in archaic societies* (2002). Building upon several studies of 'traditional' societies, Mauss argued for the importance of giving for maintaining social structures and building communities, a concept later used by Warren Hagstrom to explain how scientists' sharing of scientific results contributed to a system of reciprocity similar to the traditional one (1982).

Today, the gift economy anthropological model has been applied to the Internet and to the collaborative web in which anyone can participate in the production and sharing of free content. Once we buy a computer and we pay a provider to have access to the Internet, most online activities are free. We can use free services such as search engines, mail boxes, social networking websites, online newspapers and tourist guides. At the same time we produce content without being paid – for example, when we publish a video, share a band's record in a peer-to-peer network or write a Wikipedia entry. What allows this economy to survive and be sustainable?

Mauss' gift paradigm has been applied to the Internet in order to analyse the phenomenon of content production by thousands of individuals who choose to donate their time and abilities to projects from which they will not receive any monetary benefit (Aime and Cossetta 2010). According to this view, when we produce online we are giving – producing social bonds and community and accumulating social and symbolic power. Obviously, there are many famous examples of online collaborative production creating bonds and allowing the rise of communities not based on making a profit. The emergence of collaborative web has stimulated a wave of studies that tackle online cooperation and study it from both an institutional and political perspective.[10] Some of these communities have proven extremely efficient in creating new products or content that anyone can use for free – think about encyclopaedias (Wikipedia), free software (Linux), news (the sphere of news produced in blogs and social networking sites) or restaurant reviews (Yelp). Those projects have become famous examples of open innovation and open production that Internet users participate in without a centralised hierarchy or a wage-labour system.

Yet the gift theory taken from the Maussian tradition is not a viable general theory of online production. To highlight the differences between the gift as a 'total social fact' and the gift online is not enough: this perspective misses a crucial dimension of the mechanisms at work on the Internet. The gift, online, corresponds to new and emergent business models – still unstable and evolving, yet noticeable and relevant. The sharing culture that has grown on the Internet now involves huge numbers of people in projects of online voluntary production: the wealth, diversity and productive capacity of those communities is exceptional, and has caused much talk about a new and emerging production paradigm. But anti-utilitarianism is not an adequate criterion to understand these phenomena. There are other actors whose practices we must take into account: private companies.

Co-optation (appropriation) clashes occur between knowledge producers and companies on a daily basis. The Maussian lens is inadequate to understand what we do when we post a video on YouTube, update our Facebook profile or review a book on Amazon. Online, gift's anti-utilitarianism is unidirectional: it applies to the users who produce content without being paid, but not to the companies that transformed participation and gave into a business model. Online giving is not complementary to mercantile relationships: it is integral to and indissoluble from them. Open, horizontal, peer-to-peer production models can be interpreted as forms of community creation through giving only if, in the first place, one analyses their relationship with the dynamics of capitalist accumulation.

The contemporary digital and informational economy and its model of soft, flexible, horizontal capitalism has subsumed Mauss' gift, in a new primitive accumulation. In the twenty-first century, the gift economy is embodied in the network with its emphasis on access, participation, gratuity and sharing. It has become a new economic model and a new form of appropriation of the value produced by online cooperation. The currency of these digital potlatches[11] is not only reputation, status, sociality, community building or political power. Open source and gratuity are aspects of today's capitalist mode of production, and the rhetoric that surrounds them is often related to free market and technological advancement. On the one hand, this is nothing new if, as Armand Mattelart (2003) points out, during every new technological cycle the redeemer discourse of the information society emerges again, and the long history of the free flow of information is strictly related to deregulation and neoliberalism. For Daniel Bell (1973) the post-industrial society would have been based on cooperation and reciprocity. Yet on the other hand the contemporary digital economy has created new conditions for the exploitation of the flow of information.

Open source software corporations such as Sun Microsystem or IBM (Benkler 2006) guarantee everyone access to their codes and they sell their services, training and customisations without adopting a monopolistic management of information. This open source informational model of capitalism is presented as a crucial instrument for innovation. Other companies, such as social media websites or search engines, harvest the content produced by masses of users and sell people's personal and social data in order to create advertising revenue. Richard Barbrook's definition for this phenomenon is the 'hi-tech gift economy', an economy where most users see the Internet as a place to work but also to play, love, learn, debate and collaborate with other people without restrictions due to physical distance or copyright, and without the direct mediation of money (2000).

Barbrook underlines the impossibility for capital to completely subsume the gift economies it has to exploit and promote. Rather than prospecting either the victory of digital capital or the irrec-oncilability of capital and gift economy, he outlines the existence of a symbiotic relationship between the anarcho-communism of hackers and corporate capital. Against the utopian visions on gift economy relations as alternative to capitalist relationships, Barbrook maintains that the commodity and the gift can coexist and are not always in conflict with each other, as each form of organisation of information production and sharing do not harm or supplant the other. In cutting-edge areas of the digital economy the utopia of a hi-tech gift economy that 'heralds the end of private property' clashes with digital capital's need to privatise the gift and manage and enclose the social spaces where free and voluntary cooperation explodes. Economies based on free and open information sharing can be counterforces opposing privatisation and enclosing, and at the same time crucial components of digital capitalism's dynamics (see also Hardt and Negri 2000).

Other authors adopt a more pessimist perspective and analyse the dark side of Internet gift economies: capital's parasitism of the digital commons has been exemplified through the analysis of the exploitation of free labour by web companies that provide free content. This is 'a trait of the cultural economy at large, and an important, and yet undervalued, force in advanced capitalist societies' (Terranova 2000, p. 33) in which productive activities of cooperation and sharing are voluntarily embraced by users and at the same time exploited by companies.

Here the opposition between capital and labour is not merely based on the clash between intellectual property rights and gift economies, but is rooted in the fact that the provision of 'free labour' is fundamental to the processes of value creation in digital economies. Furthermore, incorporation, or the process of capital absorbing the fruits of underground subcultures and resistance movements, is not a mere co-optation of an independent culture by capital. Terranova sees it as an immanent process of moving collective labour that emerged outside companies into monetary flows, and thus of structuring it within production modes and business practices (2000, p. 39). This bitter critique of the much-debated liberation potential of gift economies drags the free culture movement into a fate of damnation. The open source movement is explicitly taken as an example of the free labour model when Terranova states that exploitation is endemic in the digital economy and open source, which relies on developers' free labour, is an evidence of this structural trend. In conclusion, in this portrait of participation in the dynamics of the Internet, software companies use the free labour provided by open online communities to produce and extract value. Private profit, in online business models, is the fruit of the moral obligation to share, debug, cooperate and micro-fund, which feeds hi-tech gift economies.

Citizens' participation in digital economies through information production and sharing thus can be instrumental to digital capital evolution and can become part of the processes that open up productive spaces based on non-market and peer-to-peer relationships. Both the utopian gift economy heralded by enthusiast views of the Internet's liberation potential and the dystopian expositions of massive user-exploitation somehow miss a point: that in contemporary information societies it is impossible to separate the role of sharing and participation as market tools from their role as critical and oppositional practices.

Indeed, discourses of participation, gratuity, sharing and free labour are ubiquitous in Western societies. Gilles Deleuze used the phrase 'society of control' (1992) to describe a new phase of capitalism in which disciplinary societies and modern enclosures are replaced by free-floating control and modulation on the one hand, and by people's participation, motivation and permanent training on the other. Also, in the technoscientific realm, participation has been criticised as a new form of governmentality (see for example Pestre 2008).

In genetics, one interesting example comes again from Google. The normative drive towards online participation forms the basis of the functioning of Google's genomics start-up 23andMe, founded by Silicon Valley venture capitalists and aspiring science entrepreneurs Anne Wojcicki and Linda Avey. Wojcicki is the wife of Google's founder Sergey Brin. A company that provides services such as personal DNA testing and direct-to-consumer genotyping, 23andMe has been referred to as a source of ethical dilemmas and ambiguity from the medical and public health viewpoints (see McGuire et al. 2010).

What is interesting from the perspective of studying open science politics is that 23andWe connects its customers through its social network website and urges them to share their genetic and medical information. In fact, 23andMe and other companies that provide direct-to-consumer genetic testing and whole genome sequencing provide customers with a personal genetic profile composed of extensive genome analysis. People can then use their profiles for ancestry research or for medical reasons: they get to know, with all the ambiguity inherent to these type of genetic testing, their genetic predisposition to dozens of pathologies. 23andMe customers are asked to share both this data and information about their health and medical conditions in the company's social media. According to 23andMe, 'this new approach lets *you* initiate, advise and participate in research via the Internet' (emphasis mine) and the possible results of *your* participation and sharing is to:

> eliminate the need for inefficient recruitment procedures and distribute the cost of genotyping, we believe connecting people with scientists empowers everyone to accelerate the pace of research. (23andMe, quoted in Levina 2010, p. 4)

23andMe exploits the moral obligation to share, participate and facilitate information flow in order to utilise users' data in processes and parts of its research activities that would otherwise require paid staff – such as the subjects that pharmaceutical companies hire for testing new molecules and protocols. Levina uses the notion of free labour in order to link contemporary personal genomics with social media technologies which enable, push and absorb social participation and users 'gifts' into their practices – in this case, individuals' genetic and medical data. Through its social media website and its discourses of participation and inclusion, 23andMe facilitates an active engagement with genetic research. But 'while these engagements are presented as narratives of control, freedom, and empowerment, ... citizen bioscience is enveloped in [the] "free labor" economy of the network society' (Levina 2010, p. 7).

The circulation of biological information through online communities can be created, managed and expropriated by private corporations. Users who live within the participation ideology of the web are required to share not only their personal data, as in social media websites such as Facebook or Bebo, but also biological information: once again the potlatch and the commodity are not irreconcilable, as new business models coincide with the gift economies based on the imperative to share. Biological information becomes a valuable commodity exactly when it flows through the network in order to guarantee 'freedom from disciplinary institutions through full participation in the control society' (Levina 2010, p. 5).

Furthermore, open sharing of biomedical information can allow companies to increase their ability to attract and leverage financial capital, pretty much like many web companies do (Arvidsson and Colleoni 2012). The exploitation of citizens' participation to biosciences is somehow the dark side of the increased political and scientific productivity brought about by the rise of research projects based on distributed expertise.[12] I will further explore the problem of participation in citizen bioscience in Chapter 6, when analysing the role of a citizen biology project such as the DIYbio network. We now need to focus on the relationship between hacker cultures and science ethos in order to use these concepts as tools to understand the role of a changing scientists' culture in current informational capitalism and biology innovation regimes.

3
Hackers, Rebels and Profiteers

Whoever does not adapt his manner of life to the conditions of capitalistic success must go under, or at least cannot rise.

Max Weber, 1905

From the analysis of the realm of cultural production – the production and sharing of culture in the form of online scientific information and knowledge – I will shift to the analysis of the culture of information and knowledge producers.

Scientists who decide to share information, content and knowledge for free and thus to participate in the new political economy of open science challenge our understanding of science's cultural boundaries. The Mertonian ethos of open and disinterested science was the expression of a peculiar social contract and the result of clashes around intellectual property rights that took place during the first half of the twentieth century. But new open science does not merely rely on new technological solutions and on a revival of the twentieth-century open science culture. New incentive systems and social configurations are at work, and open science has new ways of distributing benefits to those scientists who decide to share their knowledge and data openly.

However, science's communication and publishing systems have always developed in order to respond to the incentives society gives to their users. What could an updated social contract look like, then, from the viewpoint of scientists' culture? Scientists' strategies of action must enable them to participate in a system that includes private actors, corporations, foundations, citizen science projects and peer-to-peer research, as well as very different forms of information management and property.

Hacker cultures represent an important driving force for contemporary innovation regimes and are somehow an heir of scientists' culture. The interplay between this cultural system and the modern scientific ethos is contributing to the shaping of science's social contract, as it is related to the transformations that have occurred in the way science and innovation are organised,

49

institutionalised and managed. Yet hacking is the subject of a body of literature that is not always linked to more general ideas about how cultures of sharing operate in contemporary digital economies. In this chapter I try to shed some light on this relationship and introduce characters such as the rebel scientist and the scientist/entrepreneur. These two recurring figures have close relationships with hackers, and they clarify the relevance of the hacker ethic in open science.

ON SPIRITS AND IDEOLOGIES

The emergence and diffusion of commons-based peer production practices and 'hi-tech gift economies', and the participatory, cooperative turn of informational capitalism, have a cultural side. The development of new business models and corporate practices relies on, and shapes, a cultural transformation that implies normative drives towards horizontality, participation, cooperation, giving, flat hierarchies and networking. Several authors have indicated a correlation between these transformations and the emergence of new cultures that, drawing from the title of one of the most famous academic accounts of this phenomenon, have been said to be part of a 'new spirit' of capitalism.

This wave of studies deals with the ideological changes that have accompanied recent transformations in capitalism (Boltanski and Chiappello 2005). Several authors share the idea that transformations that involve people's daily practices and cultures have the power to modify, shape and drive the evolution of capitalism. But they differ when it comes to illustrating the relationship of cause and effect between cultural and material transformations, and regarding the role of those transformations in opening up new possibilities for anticapitalist and liberation struggles. Yet it is possible to draw a comparison between the vocation (the 'beruf') of scientists and their role in the evolution of contemporary capitalism.

In his classic study on the Protestant ethic and the spirit of capitalism, Max Weber argued that ethic was a form of legitimation of socio-economic structures (2003). Differing radically from Marx, he did not believe that the Calvinist ethos was an ideology produced by economic and productive conditions. According to Weber, culture is neither created by nor dependent on economic structures, and does not completely determine them. The 'spirit' exists before modern capitalism: it has appeared in places and times in which capitalist structures were still coming to light, and yet entrepreneurs' and workers' acceptance of this ethical framework was a condition for the existence of capitalism. However, this relationship is complex and multifaceted. A process of coevolution in which a pre-existing culture and a productive model can adapt to and reconfigure each other. Nevertheless, the origins of the spirit of capitalism are to be found outside and before capitalism itself – not as an individual ethos but as a social phenomena originating in groups and collective movements, 'a way of life common to whole groups of men' (p. 55).

Within contemporary sociological debate reinterpretations and renewals of the spirit of capitalism have been proposed by revisiting the relationship between the emergence of a culture shared by a social group and the evolution of capitalism, even though only some authors have explicitly addressed digital economies. These updates to the description of the capitalist ethos and professional ethics takes into account values such as flexibility, networking, horizontality, giving, cooperation and the like: values that resonate with the trans-formations of contemporary capitalism and that give people new, good reasons for devoting themselves to their work and new sets of moral justifications and normative support for their participation in the dynamics of capitalist accumulation. In *The rise of the network society*, Manuel Castells draws an incomplete yet fascinating picture of the 'spirit of informationalism'. This spirit is a common cultural code shared in diverse forms by the network enterprise. The latter is the 'idealtype' which is driving the development and dynamics of network society thanks to its new ethical structure and its cultural/institutional configuration (1996, p. 211).

The role and effects of 1960s countercultures, after the end of Fordism and industrialism, are at the centre of several analyses of the relation between culture and capitalism's evolution. Luc Boltanski and Eve Chiappello's work relies upon a study of French cadres, from the influence of 1968 critiques of capitalism to the late 1990s capitalism's attempts at renewing its ideological and organisational foundations (2005). Their analysis is echoed by David Harvey's history of neoliberalism: a phenomenon partly founded on the appropriation of 1968 ideals of individual freedom turned into a populist culture of consumerism and individual libertarianism. According to Harvey, capital had in fact the power to split off the search for social justice from the 1968 movements' rhetoric (2005).

Boltanski and Chiappello have indicated the existence of a new spirit of capitalism rooted in 1968's libertarian, hedonist and individualist values, and thus update the Weberian theoretical apparatus, giving the spirit of capitalism new roles and new origins.

According to them, capitalism now lives on critical cultures. It needs to reconfigure and adapt to them in order to be renewed and to find new ways out of the recurring impasses that block it. This mechanism allows capitalism to incorporate critiques and survive attacks. Indeed, it needs to orientate towards the common good in order to exploit committed engagement, but its own resources are not enough. In order to continuously generate its spirit, capitalism needs enemies, 'people whom it outrages and who are opposed to it' (2005, p. 27). It is from its enemies that capitalism can acknowledge and incorporate the mechanisms of justice it needs in order to demonstrate it is directed towards the common good.

According to Boltanski and Chiappello, if the capitalist system has proved infinitely more robust and stable than its detractors thought, it is also because of its very peculiar way of founding its evolution on the critiques that oppose it. Capitalism, to follow the French authors' view, 'mobilizes "already existing" things whose legitimacy is guaranteed, to which it is going to give a new twist by combining them with the exigency of capital accumulation' (2005, p. 20). Capitalism needs enemies and critiques, different accumulation paradigms, opposing cultures, and needs to actively remix and incorporate them into the new cultural frameworks adapted to its goals.

But how does the cultural dimension push people to participate actively in capitalist dynamics? According to Weber individual motivations were linked to a religious dimension, while Boltanski and Chiappello attribute a different motivation to the justification of capitalist behaviour models: the engagement in capitalist enterprise to serve the common good. Three different typologies of motivations drive a social group towards the adoption of capitalist behaviours. The first is committed engagement in the processes of accumulation as 'a source of enthusiasm, even for those who will not necessarily be the main beneficiaries of the profits that are made', (2005, p. 16) which are based on expectations of autonomy. This motivation

> is focused on the person of the bourgeois entrepreneur and the description of bourgeois values. The image of the entrepreneur, the captain of industry, the conquistador, encapsulates the heroic elements of the portrait, stressing gambles, speculation, risk, innovation. On a broader scale, for more numerous social categories the capitalist adventure is embodied in the primarily spatial or geographical liberation made possible by the development of the means of communication and wage-labour,

which allow the young to emancipate themselves from ... traditional forms of personal dependence. (2005, p. 17)

The second motivation is based on an expectation of security for themselves and their children, and the third justified in terms of the common good that can be defended against accusations of injustice. These justifications are based on a belief in the benefit of progress, the future, science and technology coupled with civic ideals that encompass 'institutional solidarity, the socialization of production, distribution and consumption, and collaboration between large firms and the state in pursuit of social justice' (2005, p. 18). Attraction and fascination, economic security and the common good. Together, these three motivations constitute 'a justificatory apparatus attuned to the concrete forms taken by capital accumulation in a given period' (2005, p. 20).

Other authors that are not related to this theoretical tradition but have addressed digital economies share several similarities with Boltanski and Chiappello's approach. Some outline the relationship between counterculture and computer culture by analysing the 1990s do-it-yourself culture and the ways in which New Left values of individual freedom and cultural dissent have been put to work (Barbrook 1998). Indeed, at the dawn of the personal computer era New Left activists stated that information wants to be free and were inspired by computer scientists who were already living within the academic gift economy. Yet Barbrook suggests that in the Internet economy, 'contrary to the ethical-aesthetics vision of the New Left, money-commodity and gift relations are not just in conflict with each other, but also coexist in symbiosis'. Unlike the authors linked to the Weberian tradition, he describes a betrayal of ideals based on ambiguity and co-optation, and not the classic view on the relations between critical cultures and capitalism's dynamics. The Californian ideology, a merciless portrait of the rise of the Internet industry, is 'a heterogeneous orthodoxy for the coming information age' soaked by hackers', baby boomers', capitalists' and countercultures' values that

promiscuously combine ... the free-wheeling spirit of the hippies and the entrepreneurial zeal of the yuppies. This amalgamation of opposites has been achieved through a profound faith in the emancipatory potential of the new information technologies. In the digital utopia, everybody will be both hip and rich. Not surprisingly, this optimistic vision of the future has been enthusiastically embraced by computer nerds, slacker students, innovative capitalists, social activists, trendy academics, futurist bureaucrats and opportunistic politicians across the USA. (Barbrook and Cameron 1996, p. 1)

The exploitation of this ideology allows capitalism to 'diversify and intensify the creative powers of human labour' by simultaneously reflecting the needs of market economics and the freedom of 'hippie artisanship', and by blurring the cultural divide between countercultural and corporate bureaucracy values. According to the missionaries of the Californian ideology, individualism, anti-bureaucracy, autonomy and do-it-yourself are the cool and hi-tech versions of the moral justifications that compose the spirits of capitalism. Computers and information technologies are meant to empower individuals, enhance personal freedom and reduce the power of the nation-state, and thus should restructure power relations in favour of utopian relations between autonomous individuals through new information and communication technologies.

For Barbrook and Cameron, the acolytes of the Californian ideology are 'McLuhanites' who claim that the government should stay off the backs of the cool and courageous entrepreneurs who drive the computer revolution. In the utopian visions sparked by this ideology, technical solutions and the free market will replace bureaucracies and prove to be more efficient. Government intervention is considered an interference with the emergent properties of the new economic and technological forces that, we are told, represent both the future and today's embodiment of the laws of nature.

Other studies of hackers and their values underline that hackers are part of a new way of organising labour and production. For example, in their survey on the values of free and open source software developers, Mikkonen, Vadén and Vainio (2007) maintain that in corporate environments traditional hacker values of freedom and sharing have much less importance, as developers may not be interested in issues of copyright and the free sharing of software. Anticapitalist values are seldom present and the old Protestant ethic of work is 'striking back': in communities that produce free or open source software under a corporate umbrella, traditional 'Weberian' values of labour organisation are still in place. Yet this portrayal of the hacker is different from the one that characterises other studies focused on the hacker ethic as a part of a series of professional ethics that have been interpreted as renewals of the spirit of capitalism, adapted to the transformations of the information society. According to Pekka Himanen (Himanen and Torvalds 2001), within the information society the hacker ethic challenges the Protestant work ethic, and it is 'an alternative spirit' characterised by passion, freedom, social worth, openness, activity, caring and creativity.

These authors, whose approach to capitalism's evolution differ significantly on several issues, share a common point of view. According to them, critical cultures can be co-opted and incorporated into corporate strategies, thus actively shaping the development and transformation of today's capitalism. My goal, then, is to understand how the new culture, whose emergence I point out in this work, is organised, where its historical roots are, and finally how it participates in the shaping of contemporary biology.

This culture, following the viewpoint that emerges from the work of the new spirit of capitalism theorists, actively contributes to the production of the capitalist society in which it flourishes. Yet its roots in critical movements, and anti-corporate and anti-privatisation practices, urge us to rethink the unidirectionality of this perspective. If, as Boltanski and Chiappello argue, critical cultures are vital for the evolution of capitalism towards new ways of organising production and labour, it is also because of the threats posed by those critiques that capitalism is forced to move permanently and transform itself. Yet exactly because of hacker cultures' complexity, the study of hacking might provide fresh ideas that can contribute to these general frameworks on the evolution of capitalism, thus allowing us to rethink critical cultures' role in the birth and renewal of new spirits of capitalism. This is not only because they force us to take into account cultures and practices not directly related to the post-1968 countercultures, but also because they point out the need to deepen our understanding of the agency these movements and cultures can have *on* capitalism beyond the dimensions of co-optation and recuperation *by* capitalism.

Two characters that I chose to be representative of the cultural transformations I am highlighting might help us to understand the relationship between hackers and scientists. These are the rebel and the profiteer, and are useful tools we can use to better understand contemporary bioscientists and the relationship between their culture, the world of ICT and software production, and the hacker history and myth.

HACKERS

Who are hackers, then, and what is their ethic? Let's begin from a rather crystallised version of it. The hacker ethic is a contemporary set of values related to innovation and research that participate in the development of contemporary capitalism. For the sake of my analysis, I consider the precepts of a hacker ethic as an analytical tool to understand the new public ethos embodied by the open science culture I highlight in this study.

The hacker is an innovator who has never faced the problem of separation between industry and academy. According to the most widespread mythologies, hackers were born at the Massachusetts Institute of Technology (MIT). Their ethic was formalised for the first time in 1984 by Stephen Levy in his book *Hackers: Heroes of the computer revolution* (2010a) and was considered as a direct heir of the twentieth-century academic scientist's ethos. Yet it became detached, becoming more multiform and more attuned to the economic dynamics of the software world, made of start-ups, people escaping from the academy, corporate networks, garages and computer science departments.[1]

The communal ethos of hackers, coupled with values such as the free sharing of information and knowledge, and peer recognition, resonates with the behaviour of modern scientists' communities. A great example of biologists organised as *ante litteram* hackers is the history of the drosophila experimental group in the post-World War I United States. These geneticists used to work collectively and were discouraged from transforming lines of work into their personal domains. In that research environment, everyone 'meddled in everyone's work all the time, swapping mutants, ideas and craft lore' (Kohler 1994, p. 91; see also Kelty 2008). The drosophila community was driven by strategies of improvisation, and free exchange was a fundamental feature of its productive economy as well as its moral economy: a crucial part of the professional identity of the members of the group.

Yet it is important to recognise that the hacker ethic is not a stable or institutionalised concept or set of norms. According to Stephen Levy, for the first generation of hackers its precepts 'were not so much debated and discussed as silently agreed upon. No manifestos were issued. No missionaries tried to gather converts' (2010a, p. 27). Most hackers state explicitly that they do not recognise themselves in any of the main accounts of hacker ethics, or in any of the several hacker manifestos that have been issued over the years. Thus the main narrations and studies on hackers, their ethic and their history decidedly diverge when it comes to scope, object and results.

Hacking is in fact a very diverse and heterogeneous phenomenon. Its degree of plurality is incredibly wide, as it includes many practices that all vary in different ways, such as breaking into closed systems, stealing data, coding and sharing free software, implementing hands-on approaches to technology, protecting online users' privacy, enacting transparent tools and platforms, and developing critical thought about technologies. This wealth of practices goes hand in hand with an extreme political diversity and with a high degree of plurality and heterogeneity (Coleman and Golub 2008; Maxigas 2012).

For Christopher Kelty, hackers are difficult to define, not because of their multiform complexity, but precisely because what they do is introduce new technopolitical entities into the world. Thus, defining who and what hackers are is an open-ended enterprise. Furthermore, hackers ('geeks' according to Kelty's definition) constitute a public that is recursively focused on producing and reproducing the technical and legal conditions for its own existence, as well as the cultural and discursive ties that make it a community (Kelty 2008). But now let's introduce this complexity with one of the classic accounts of hacking.

Levy describes an ethic composed by 'inquisitive intensity, skepticism toward bureaucracy, openness to creativity, unselfishness in sharing accomplishments, urge to make improvements, and desire to build' (2010a, p. 37). In this respect, we can highlight some common features. The hacker, born at the MIT Artificial Intelligence Lab in the late 1950s and early 1960s, grew under the influence of the American countercultural movements of the 1970s (Turner 2006; on the links with biology see Vettel 2008) and is not just an independent, curiosity-driven innovator with a proactive attitude towards technology and committed to information sharing. The hacker is also a heretic, a rebel against institutions and bureaucracy, a hedonist who works for fun and to make the world a better place. And yes, the hacker is also a resource ready to be sold to venture capital.

Several studies which address the discourses of hacker communities underline their ambivalence in regard to their relationship with capital. According to Kirsty Best 'hackers are not only a fringe element, but an integral part of the dominant social ordering of technology. The challenges they make contest and undermine technological systems from within, by exposing gaps and holes in the fabric of technology' (2003b, p. 256). Some authors explicitly describe the hacker ethic as a new or alternative spirit of capitalism – the best known example is Pekka Himanen (Himanen and Torvalds 2001, see above).

On the contrary, authors openly coming from the hacker world depict this ethic as a tool for resisting digital capitalism. For Ippolita, the collective hacker norms of behaviour are ideal ways of relating to technology in an active and non-authoritarian way:

> Let's idealize the hacker: passionate study, self education outside the market, curiosity and exchange with referential communities, broad and variegated networks of relationships. The hacker does not settle for tales, whether truthful or not, but needs to spot the source, touch the fount, the origin. Put his/her hands on it. (Ippolita 2005, p. 106)

Even though Levy's book presents the hacker ethic in a static way, it can serve as an introduction to the inextricable intertwining of these different facets of hacker culture. In fact it is a celebration of hacking as embodied by the very first group of hero adventurers who made the wonders of the 'computer revolution' possible. Drawing on the history of the first generation of hardware hackers, the kids from the early 1960s who worked on the TX-0 mainframe in building 26 at MIT, Levy describes the hacking culture of the very old school, but his principles have been quoted *ad infinitum* by new generation hackers. His version of the hacker ethic is as follows:

- *Access to computers – and anything that might teach you something about the way the world works – should be unlimited and total. Always yield to the Hands-On Imperative!* When the Midnight Requisitioning Committee needed a set of diodes or some extra relays to build some new feature into The System, a few people would wait until dark and find their way into the places where those things were kept. None of the hackers, who were as a rule scrupulously honest in other matters, seemed to equate this with 'stealing.' A wilful blindness.
- *All information should be free.* If you don't have access to the information you need to improve things, how can you fix them? In the hacker viewpoint, any system could benefit from an open flow of information.

- *Mistrust authority – promote decentralisation.* The last thing you need is a bureaucracy. Bureaucracies, whether corporate, government, or university, are flawed systems, dangerous in that they cannot accommodate the exploratory impulse of true hackers.
- *Hackers should be judged by their hacking, not bogus criteria such as degrees, age, race, or position.* Hackers care less about someone's superficial characteristics than they do about their potential to advance the general state of hacking, to create new programs to admire, to talk about that new feature in the system.
- *You can create art and beauty on a computer.* To hackers the code of the programme holds a beauty of its own.
- *Computers can change your life for the better.* Surely everyone could benefit from a world based on the Hacker Ethic. If everyone could interact with computers with the same innocent, productive, creative impulse that hackers do, the Hacker Ethic might spread through society like a benevolent ripple, and computers would indeed change the world for the better.[2]

Obviously, one of the main ingredients – if not the primary one – of the hacker myths is the emphasis on 'active access to information' promoted and pursued by hackers (Best 2003b). In the 1970s, Captain Crunch was one of the first *phreakers*, hackers able to break into the American telephone network, and is still today a mythical figure of the hacker iconography. He did not act for money, but for the eagerness to know the codes managing the network, which he revealed to everyone, along with the tricks to use them. Crunch would break into phone systems to learn and explore: 'I'm learning about a system. The phone company is a System. Do you understand? If I do what I do, it is only to explore a System. That's my bag. The phone company is nothing but a computer' (quoted in Levy 2010a, p. 254).

Information is good per se and cracking a code or accessing a system are the hacker's goals. Phone hacking became an activity characterised by 'devotion to technical expertise irrespective of professional affiliation; the intrepid exploration of a network; the discovery of knowledge; the free sharing of discoveries with the priesthood of experts' (Johns 2009a, p. 466).

In order to become 'crackable', DNA also had to be transformed into pure code. While this change opened up life sciences to new forms of citizens' distributed participation, it also triggered new appropriation possibilities. The sequencing of the human genome is an informational milestone in the history of biotechnology and bioinformatics. Although the informational DNA metaphors (the book of life, the code) date back to the origin of modern genetics, many scholars have dealt with the analysis of the role played by the Human Genome Project and by the Celera Genomics of Craig Venter in establishing a model of genetics based on information technologies and in its impact on the practices linked to intellectual property and to the size of the contemporary genomics market (Hilgartner 1995; Kay 2000).

Kay maintained that 'genomic textuality' had become crucial not only for the scientific development of genomics, but also for its commercial development. Other scholars have analysed the economic transformations linked to the post-genomic era and the information flows marking it, arguing that it is a new form of biocapitalism where technological and economic links between contemporary genetics and ICT have become stronger (Franklin and Lock 2003; Sunder Rajan 2006). It is also important to stress the deep role, both from the epistemic and socio-economic point of view, played by the 'cybernetic turn': a turn towards the translation of genes and bodies into code, where informational pattern is privileged over materiality (Hayles 1999; Waldby 2000) and the incorporation of information on a biological substrate is only a contingent event. Using Haraway's words (1991, p. 164):

> communications sciences and modern biologies are constructed by a common move – the translation of the world into a problem of coding, a search for a common language in which all resistance to instrumental control disappears and all heterogeneity can be submitted to disassembly, reassembly, investment, and exchange.

Furthermore, hackers feel a deep hatred against code restrictions: they do not tolerate the prohibitions that prevent people from accessing the information that makes up the programme instruction. Sharing is also one of the most important commandments of the hacker ethos. Richard Stallman is the hacker (the last of the true hackers, according to the Levy's book) that founded the free software movement by writing the operating system GNU – the basis of Linux – and the GNU Public License, precursor of the more famous CC licenses. In 1984 Stallman resigned from MIT over a controversy on the free sharing of software code, and with a sophisticated legal 'hack' that gave birth to the free software movement he opened up a new space for both collective and corporate action (Kelty 2008).

For hackers, data enclosure and privatisation might even be considered crimes. During the 1970s, Bill Gates was in contact with the Homebrew Computer Club based in Silicon Valley, then the new epicentre of the hacker movement. The club was a site for hackers to share information, knowledge, tricks and code: something Gates was producing. Bill Gates became the 'bad boy' of software in part because of his infamous 'Open letter to hobbyists', published in the *Homebrew Computer Club Newsletter* in January 1976, in which he complained about the free circulation of software among the hacker community. Hobbyists were illegally copying and distributing his (and Paul Allen's) Altair Basic (which, thanks to this form of piracy, became the *de facto* standard, to Gates' pleasure). Yet hackers' reactions were negative: Gates received between three and four hundred letters, and most of them were intensely negative. Stir and disdain shook the hacker community after the publication of the letter, an event later known as 'the software flap' among hacker communities. In his letter, Gates went quickly to the heart of the matter:

> Why is this? ... As the majority of hobbyists must be aware, most of you steal your software. ... Hardware must be paid for, but software is something to share. Who cares if the people who worked on it get paid? (Gates 1976, p. 1)

Besides the focus on pure information and open access, hacker ethics are multiform and radically ambivalent. The hacker is not only an independent, curiosity-driven innovator, dedicated to sharing his/her knowledge, but also a heretic, a rebel against institutions and a resource ready to be sold to venture capital. Some of these characteristics of the hacker are mirrored in the public image and history of modern scientists, who can be rebels and profiteers as well.

REBELS

Autonomy is one of the important frameworks that define modern scientists. Michael Polanyi famously stated that the autonomy of scientists had an epistemological motivation, or it was necessary for science to be more efficient: 'any attempt at guiding scientific research towards a purpose other than its own is an attempt to deflect it from the advancement of science' (1962). Governmental and corporate planning was rejected 'as antithetical to the very idea of science', as Shapin puts it (2008, p. 197). An even deeper rejection of authority and planning resided in the rebel, iconoclast, maverick and heretic scientist as a classic element of the narrations on modern science and biology.

Revolutionary science is the engine of scientific advancement in Thomas Kuhn's *The structure of scientific revolutions* (1996). Kuhn introduced sociology in the epistemological approach and attributed paradigm shifts to the efforts of young scientists towards the imposition of new ideas over established ones. In a non-academic book, the physicist Freeman Dyson (2006) tells the stories of rebels such as Isaac Newton, Robert Oppenheimer, Richard Feynman and Edward Teller, scientists who built their careers on the willingness to not abide by the rules of the status quo. In a collection edited by Oren Harman and Michael Dietrich several science historians analyse different figures of *Rebels, mavericks, and heretics in biology*. Iconoclast scientists embody different ways of challenging the status quo, but, as Harman and Dietrich put it, they

> are living testaments to the irreverent existence of free will (and thought) in the face of what might seem, to their more conventional counterparts, necessities or truths in no need of being challenged. (2008, p. 9)

In some cases, rebellion becomes part of the researcher's public image, which then becomes a full-blown 'public myth'. This is the case of the famous geneticist Barbara McClintock, who combined public iconoclasm and private rebellion (Comfort 2008; Keller 1983). Nevertheless, often the iconoclast becomes an icon when roles switch and the rebel gets full recognition from the scientific community or other communities (in McClintock's case feminist historians and philosophers, as well as the Nobel Prize committee). Obviously, 'even rebels need a framework within which their rebellion will make sense' (Segerstrale 2008, p. 297) and rebels without a cause are rarely able to find their way to the top of contemporary science.

Richard Lewontin lists elements that we need to take into account when we analyse the success of rebels: for example, scientists need to perform public communication and to struggle for employment, promotions and grants. In a more general sense, their acts of rebellion are directed against social or political power. Indeed, in order to understand these rebellions, one should put them in the context of the social and economic structure that enable people to maintain and propagate their thoughts and their influence. According to Lewontin, 'breakers of idols are not smashing mere representations of others' gods but destroying potential rallying points for the collective activity of other sects', thus answering to a social and political, rather than epistemological, need (2008, p. 372).

Although rebellion is often 'a retrospective self-description' from scientists themselves, 'being a rebel even appears to be a strategy to attain and keep power' (Morange 2008). After all, Polanyi himself argued that scientific dissent is often not directed towards scientific institutions but rather against the interference of other types of authorities, and other authors have argued that absolute autonomy in an academic context is a myth, for issues of funding, politics and relations with private firms always condition the scientists' activity (see Krimsky 2006; Shapin 2008).

Harman and Dietrich also emphasise that the rebel scientist can appear both inside and outside the most important scientific institution: the university. Not all rebels must work outside the academy, yet often a rebel scientist has to make a break with their institution and the authority of their peers. The British biochemist Peter Mitchell could not conform with university life and chose to work in his personal research institute, Glynn, and to publish his books via an independent publisher he founded, Grey Books. In some cases rebellion against universities was directed against patenting. In the 1950s the American mathematician Norbert Wiener portrayed himself as a rebel by declining government funds and laying aside research 'to concentrate on exposing what he saw as the corruption of science by intellectual property' – namely that the rise of corporate science and the diffusion of patents were impeding the flow of information (Johns 2009a, p. 425).

Among biologists, for example, primatologist Thelma Rowell used to maintain explicitly that she had 'always [been] taught to question authority: the more authoritarian it is, the more you question it' (Despret 2008, p. 351). The evolutionary biologist William Hamilton

> disliked authority, hierarchy, taboos, organized piety, and the growing dependence of science on profit-seeking industry. He wanted open discussion and disliked suppression of truth. He disliked political correctness and ... also liked breaking rules, at least in small ways, and liked shocking people's beliefs. (Segerstrale 2008, p. 296)

Despite these beliefs, Hamilton spent his entire life looking for sponsors to fund his research projects.

Pierre Bourdieu (2004, p. 63), while referring to epistemic (and not institutional) revolutions, highlighted that the revolutionary scientist does not only head towards a scientific victory: there's more at stake. Scientists are sometimes willing to change the rules of the game: 'revolutionaries, rather than simply playing within the limits of the game as it is, with its objective principles of price formation, transform the game and the principles of price formation'. Thus, the struggles in the scientific field are ones in which 'the dominant players are those who manage to impose the definition of science that says that the most accomplished realization of science consists in having, being and doing what they have, are and do' (Bourdieu 2004, p. 63).

To highlight a more concrete case, Harman and Dietrich (2008, p. 18) conclude their introduction with a perspective on rebellion not from the epistemological viewpoint, but the socio-economic one: in the twenty-first century, new and heterodox ideas could come from highly original and rebellious minds capable of tweaking biology's funding system, the online publishing system or the relations between university and industry. In order to revolutionise science, tomorrow's genial intellects will need to change the socio-economic structures of life sciences research rather than simply improve existing knowledge with new revolutionary ideas.

Hackers, with their search for new, heretic solutions and their distrust for authority, centralised bureaucracies and mainframe computers, are certainly in debt to the tradition of the rebel scientist. If rebel scientists revolt against academic bureaucracies or military command over science, the first generation of hackers struggled against IBM and mainframe computers which were not hackable – for example, the IBM 704 computer on the first floor of building 26 at MIT, a computer hackers used to call Hulking Giant. A Hulking Giant was a huge, slow, non-hackable computer, 'the inevitably warped outcome of Outside World bureaucracy' (Levy 2010a, p. 78). These computers were managed by what hackers used to call a 'priesthood', and were difficult to access unless one was prepared to deal with the old-fashioned bureaucracy which managed them – people in white lab-coats who were in charge of punching cards, and pressing buttons and switches. The privileged priests who could submit data to the machine and interpret its answers engaged in a sort of a ritual with their acolytes, who were not granted direct access to the mainframe:

Acolyte: Oh machine, would you accept my offer of information so you may run my program and perhaps give me a computation? Priest (on behalf of the machine): We will try. We promise nothing. (Levy 2010a, p. 5)

A similar priesthood, a 'scientific fraternity', defends science's autonomy from external impositions according to Merton. But the hackers' quest for autonomy is deeper. If IBM had its way, according to hackers, the world would be slow, centralised and bureaucratic, and 'only the most privileged of priests would be permitted to actually interact with the computer' – people who 'could never understand the obvious superiority of a decentralized system'. But on the ninth floor of MIT building 26, the floor where hackers were free to experiment with computers, nobody needed to notify superiors or fill out forms to do 'the right thing': hackers had 'no need to get a requisition form. ... Hackers had power. So it was natural to distrust any force that might try to limit the extent of that power' (Levy 2010a, pp. 30–1).

Later, Microsoft replaced IBM as the enemy of decentralised and open cooperation and innovation. In *The cathedral and the bazaar*, Eric Raymond (2001) contrasts the cooperative bazaar model of the open software initiative with the closed and hierarchical cathedral of the Microsoft organisation. By keeping its information proprietary, Microsoft obfuscates users' direct relationships with technology. Indeed the hacker often pursues knowledge in a way that is independent from hierarchies and institutions. The only acknowledgment he/she looks for comes from his/her results: to crack a code is a goal in itself, and to prove that your hack works is the only thing you need to validate your work. Hackers want to write good code, not to publish peer reviewed research papers, and they often value charismatic authority over formal and bureaucratic reward systems (O'Neil 2009).

This anti-bureaucracy attitude was quickly directed not only against Hulking Giants but against corporate and nation-state monopoly in general (Best 2003b; Johns 2009a; Levy 2010a). In fact, the history of computers and hacking has other noble ancestors that explain hackers' rebel roots: radical social movements and 1970s countercultures. A new generation of hackers was born in the San Francisco Bay Area around 1968, and its goal was to 'bring computers to the people' (Levy 2010a). These hackers worked in close relationship with the countercultural movements of Berkeley, the free-speech movement, anti-war, anti-nuke movements and so on. Groups such as the People's Computer Company (PCC) wanted to 'dissipate the aura of elitism, and even mysticism, that surrounds the world of technology' and the first-generation hackers ('Jesuits!' according to the Bay Area hacker Lee Felsenstein)[3] with the dream that 'access to terminals was going to link people together with unheard-of efficiency and ultimately change the world' (Levy 2010a, pp. 162–5).

Microprocessors – and thus the birth of the personal computer – were going to 'eliminate the Computer Priesthood once and for all' (Levy 2010a, p. 187). The principle of active access to information 'becomes translated into an expanded principle of more generalised (and recognizably democratic) fights for access, whether in response to anti-democratic practices of nation-states or commercial entities' (Best 2003b, p. 273). The title of Fred Turner's book, *From counterculture to cyberculture* (2006), indicates this transition. Turner traces the roots of cyberculture back to the 1970s, when the personal computer revolution had grown directly out of the counterculture. Turner highlights how, for the figures who bridged the New Left and computer culture, the 1980s cyberculture was in debt to a peculiar political underground:

> Bay area computer programmers had imbibed the countercultural ideals of decentralization and personalization, along with a keen sense of information's transformative potential, and had built those into a new kind of machine. (p. 103)

The rebel side of computer culture was soon used as an explicit framework for marketing. In a famous 1984 TV spot for the new Macintosh, Apple depicted computers 'as devices one could use to tear down bureaucracies and achieve individual intellectual freedom' (Turner 2006, p. 103).

The origins of the biotechnology industry are also partially rooted in post-1968 countercultures and Californian social movements. Eric Vettel reconstructs the history of Cetus, one of the first biotech companies created in the San Francisco Bay Area at the beginning of the 1970s. Vettel highlights how, 'whether they conducted experiments, published articles in scholarly journals, or delivered papers at scientific conferences, Cetus employees continued to participate in a peer society that celebrated the most professional aspects of academic research' (2008, p. 204).

Furthermore, thanks to the huge efforts made by social movements and students to change the direction of the then emerging biotech industry, working in the biotechnology industry was perceived as egalitarian and humanitarian. Some researchers, for example, had problems in dealing with their bosses: the deference they needed to exhibit clashed against the anarchist counterculture of Californian campuses. Indeed, explicit issues of participation and democracy are at stake in hacker and computer culture. Several authors have outlined that political outcomes can develop from these cultures.[4]

On the one hand, we are talking about a critical culture that clashes against the development of neoliberal capitalism and corporate power. For example, the politicised side of the hacker movement has an explicit epicentre in Italy, where every year since 1998 the national hackmeeting is held in social centres and squats and combines hacking and social activism – hacktivism.[5] Yet according to several authors, as we have seen, hacking is essentially ambivalent in its political orientation. This doesn't mean it has two separate and opposite facets, but rather that it is not possible to demarcate clearly the separation between alternative, critical and radical practices and the role of hacking in the evolution of corporate computing and hi-tech gift economies.

PROFITEERS

Even though capitalist endeavours and the values of hackers often conflict, the hacker ethic is ambivalent with regard to profit and entrepreneurship. In November 1984 Stewart Brand, the editor of the *Whole Earth Catalog*, organised an important hacker conference in the San Francisco Bay Area. Conversations were dominated by two main issues: 'the definition of a hacker ethic and the description of emerging business forms in the computer industry' (Turner 2006, p. 105).

Gabriella Coleman's ethnographic work on hacker and open source communities highlights how hackers' discourses embody liberal values such as free speech, giving birth to a form of 'political agnosticism' or 'multiple morality'. In Coleman's perspective, FLOSS hackers have given code a political neutrality made material through copyleft licenses. Thus the very ambiguous meaning of the 'free' of free software includes ideals such as 'individual autonomy, self-development, and a value-free marketplace for the expression of ideas' (Coleman 2004, p. 510).

Yet very different actors can mobilise free software meaning and interpret it according to their opposite needs: IBM adopts it as part of its neoliberal language, while anti-corporate media such as

Indymedia find in it subversive potentialities. Through hacker ethic and FLOSS practices, some can celebrate the cult of the individual while others may celebrate the collective (Jesiek 2003). Even media piracy can become a business force, and its links with libertarian ideals of distributed creativity and laissez-faire suggest that in some cases piracy's moral philosophy can lead 'not to Stewart Brand and ultimately John Stuart Mill, but to Oliver Smedley and Ronald Coase' (Johns 2009b, p. 56).

Thus there is a connection between the hacker ethic and profit – one that echoes the connection between scientists' culture and profit. Recent work in the social history of science have highlighted how the enduring image of uninterested scientists is simplistic – for instance, in money and economic matters. Disinterestedness is one of Merton's norms, yet several authors have criticised the possibility of adhering to this norm given the reality of scientists' everyday life and work. Hackers certainly do make money, but should scientists make money as well?

Of course, before the establishment of twentieth-century, publicly funded academic research, the 'man of science' became a publicly recognised figure precisely because of his role as 'a gatekeeper into the commercialization of creativity in industrial society' (Johns 2009a, p. 259). Yet making money seems to not be part of modern academic scientists' public culture, or at least there is tension between Mertonian norms and industrial counter-norms (Eisenberg 2006; Hackett 1990). As maintained by Eisenberg, 'even as their research goals and appropriation strategies have sometimes converged, academic scientists have struggled to define their norms and practices so as to distinguish their enterprise from that of their profit-seeking rival' (2006, p. 1029).

Lately, though, more researchers seem to display hybrid orientations and to exploit today's 'fuzzy boundaries' between science and business in order to defend and negotiate their positions. Subtle resistance against the commercial ethos can appear, but in this complex and fluid picture the adherence to the traditional norms of science and the entrepreneurial models of corporate research coexist. While some scientists can resist the invasion of commercial science, others 'partake in the realms of both science and business', showing no signs of any ethical problem (Lam 2010, p. 309).

In between these two poles there is a range of intertwining between disinterestedness and business. When scientists participate in patenting practices as a normal part of their professional life, they may perceive it as problematic and need to refer to more traditional values. Or, as Packer and Webster put it, 'they have to map it onto their more central activity as professional scientists' in order to conserve ties with their academic socio-technical competencies (1996, p. 450). In fact, the incorporation of patents in the scientific cycle of credit was accompanied by an active role of scientific communities and leaders that used patents to reshape the dynamics of credit in scientific careers (Murray 2011).

While open sharing practices were never limited to academic science, secrecy, patents and other forms of enclosure were not uncommon in university research. Both systems of managing information and knowledge were broadly used by industrial and academic actors during the twentieth century. This sharp boundary between academic culture, driven by Mertonian norms of disinterestedness and profit-driven corporate scientific culture, has never been absolute. The infusion of entrepreneurial values brought by increased privatisation during the last decades of the twentieth century was acting on a substrate where the two cultures were overlapping.

First of all, the commitment to the free sharing of knowledge depends on the incentives scientists follow, and so the idea of science as public and disinterested knowledge is a social expedient. Richard Barbrook insists on economic aspects when he maintains that 'in science, the opposition between giving as a form of socialising labour, and commodity was never real' (1998). In a sense he takes Hagstrom's viewpoint to an extreme. Warren Hagstrom maintained that the gift exchange within scientific communities had a specific organisational role so that it was simply functional to scientists' careers to adhere to the gift-giving principle, as it serves the peculiar interest researchers have in publicising their work in order to accumulate social capital and obtain career advancements (1982). Commenting on Hagstrom's work, Pierre Bourdieu depicts a scientific field in which 'the pressure of external demands threatens the disinterestedness of scientists or, more precisely, the specific interest in disinterestedness' (2004, p. 52). In this sense, disinterestedness is an important part of scientists' culture and informs the paradigm of gift exchange. Social profit and scientific (symbolic) profit, are inseparable and academic norms are part of this intertwining:

> every scientific choice – the area of research, the method used, the place of publication, the choice ... of rapid publication of partially verified findings – is also a social strategy of investment

oriented towards maximization of the specific, inseparably social and scientific profit offered by the field. (p. 59)

However, a deeper level of complexity needs to be added to this picture. In the scientific field the separation between the ethos of disinterest of the Republic of Science and the profit-making, entrepreneurial ethos of the Realm of Technology was never entirely true, and the accumulation of symbolic capital is often secondary to economic reasons. According to Adrian Johns, 'proprietary science might be as genuine as nonproprietary, depending on where one worked' (2009a, p. 405). In this view, the discussions over the nature and conditions of entrepreneurship can become the place where the very idea of science is at stake. Even in the mid-twentieth century, decades before the rise of research commodification and academic capitalism, the separation between the practices of the academic scientist and those of his/her profit-seeking antagonist – the industrial scientist – was not complete.

Steven Shapin (2008) addressed the relationship between industrial and academic science by reconstructing the contradictions that, in the early twentieth century, opposed 'the view from the tower' (the normative accounts of scientists' ethos written by Mertonian sociology) and 'the view from the managers' (a body of studies by organisation sociologists who worked for companies with R&D sectors). According to Shapin, the ivory tower of science never existed in the way it was depicted during the twentieth century, and the complex reconfigurations of science institutions towards research commodification were somehow part of a longer process. The worlds of academic science (driven by the search for the truth) and industrial science (driven by the search for profit) needed different public images, but at the same time shared several characteristics.

After World War II, publishing practices and intellectual property policies already depended on an ecological business model that, even in industrial settings, was becoming more and more complicated. The industrial scientist was at times able to publish results in journals, and to give talks at conferences, at meetings of scientific societies and to panels of peers. Patents were perceived to be not only tools for locking up ideas but also for communicating results, and within the faster fields of innovation the most important thing was not to maintain secrecy but to be able to stay one step ahead of competitors.

At the same time, universities were partially embracing patenting policies before the turn to academic capitalism in the 1980s. The 'managerial ethos' imposed on American – and increasingly on European – universities during the last three decades is part of informational capitalism's need to embrace new ways of managing scientists' contributions to technological development. Yet open science practices such as data sharing and open publishing re-emerged in the corporate sector. Over the history of science, wars against patent laws have been fought in turn by both free-trade advocates and antagonists of capitalist relationships. In the mid-nineteenth century, British patent laws were nearly abolished outright in the name of laissez-faire principles, and inventors were framed as 'brains' fighting against 'capital' (Johns 2009a, p. 278). However, as we saw in Chapter 1, in the twentieth century inventors had to be protected against intellectual property in the name of a type of open science that informed the evolution of capitalism itself (p. 422). So, when he writes about the professed altruism or moral virtues of scientists, Shapin is underlining the importance of personal reputation for people who deal with the 'radical uncertainty' of the technoscientific enterprise in a world where speaking of nature and technology means speaking on behalf of the future as well. The book ends – not coincidentally – with the description of a sunny day in San Diego when 200 scientists, biotech and high-tech entrepreneurs, venture capitalists, intellectual property lawyers and other path-breaking species of the knowledge ecology met and networked by the beach.

Hacking is always a subject of passion, drive and pleasure. As I have discussed elsewhere, the hacker movement is often associated with a discourse that subsumes work as passion (Levy 2010a, p. 270). As for science, fun and passion were indicated by 1960s sociologists such as Lewis Feuer (1963) who, in opposition to the Mertonian ideal of science, maintained that one of the main motivations of modern science is essentially hedonistic. Shapin persists with this point, highlighting the relationship between 'fun and funds' (2008, p. 217) and giving the example of very famous scientists such as James Watson, Richard Feynman, Kari Mullis and Craig Venter. Sure enough, hedonism got along very well with the willingness to make profits and cooperate with private corporations. After all, fun is one of the categories at work in the Silicon Valley's technology heavens. Google's relationship with the Burning Man Festival has been described by Steve Turner as providing the company's engineers with a 'cultural infrastructure' that, through fun and non-profit practices, shapes and legitimates the collaborative processes at work within the company (2009).

Furthermore, as we have seen, firms are often willing to put to work the form of socialising knowledge we see as typical of academic science. These companies may 'sometimes opt to freely disclose inventions that are patentable' even when they seem 'vulnerable to wasteful disruptions' (David 2003, pp. 8–9). In fact, making business, for hackers, is not sanctioned unless it means betraying the openness ethic – as in the case of the 'Letter to hobbyists' Bill Gates sent to the Homebrew Computer Club.

But sharing is not opposed to business, even though it can be difficult to conciliate openness and profit. When, for several second-generation hackers, going into business became 'the right thing', all of a sudden they had to face the fact that they had secrets to keep. But in the end, 'the bulk of these hackers fully integrated their skills within the capitalist enterprise system and the burgeoning information economy' (Levy 2010a, p. 266), and the same happened to the Silicon Valley generation that gave birth to the personal computer revolution. According to Apple legend, in order to collect the capital needed to start the company Steve Jobs sold his Volkswagen bus and Steve Wozniak his HP calculator. Hackers had to pay a price though, and this was somehow a re-elaboration of their ethic. According to Levy, who is eager to represent hacker ethics in perhaps a more static way than anthropologists with experience of today's hacking complexity, 'the Hacker Ethic became perhaps less pure, an inevitable result of its conflicts with the values of the outside world' (p. 451). From an opposite viewpoint, the price was the integration of the hacker ethic into corporate culture.

The interplay between commercial and non-commercial, public and private, autonomous and authoritarian, or profit and non-profit, are thus vital to describing the context in which open science operates. One of the points authors who have worked on the political significance of hacking and free software often make is that these practices are part of bigger changes in the relationship between innovation, knowledge and power. Best argues that 'the balance of power–knowledge has shifted in favor of this new, resistant, knowledge set – and that the real reason for the widespread fear of hackers is that they have outsmarted the traditional authorities in societies' (2003b, p. 267).

This does not merely mean that these practices are more effective with regards to innovation. They might often be, but another important factor at stake is their role in developing new social configurations and thus in sustaining new innovation regimes. Hackers, rebels and profiteers are cardinal points for deciphering the evolution of open biology politics that I will describe in the following chapters: the open access turn of Craig Venter, the symbol of entrepreneurial science; the rebellion of Ilaria Capua against bureaucracies and their closed data-sharing policies; the network of citizen biologists DIYbio and their attempt at translating the hacker tradition into biology.

The complexity of hacker cultures reveals some similarities, but also different ways of participating in the transformations linked to the emergence of today's open biology. These cases exemplify some of the crucial shifts biology is going through, in which the stage is taken by transformations in the peer review system, public participation in the scientific enterprise, commodification of academic biology and by the crisis of traditional scientific institutions and the rise of new social spaces for biological research.

4
Sailing and Sequencing the Genome Seas

A man who dares to waste one hour of time has not discovered the value of life.

Charles Darwin, 1836

The scientific adventure is one of discovery, of cracking nature's secrets, of sharing discovery's results. Craig Venter embarked on a scientific adventure that led him to sail the waters of the world's oceans in a search for unknown genomic resources. While sailing, surfing and swimming, Venter also undertook a voyage through databases and scientific journals, and discovered both new genes and possible new configurations of the research enterprise. In Venter's case, adventure was based on the premise that a scientist can be independent from science's institutions, share information and knowledge, have fun and make money in the process. The *Sorcerer II* is the highly mediatised and spectacular research vessel operated by the J. Craig Venter Institute (JCVI) that circumnavigated the Earth between 2003 and 2006 to collect, sequence and classify marine microbial genomes in order to build a massive metagenomics database, the data from which could be used for synthetic biology projects.[1]

Craig Venter is an American biologist who became famous in the 1990s for his role in the race to sequence the first human genome. With his Celera Genomics he challenged the public consortium Human Genome Project (HGP) after a controversy related to gene patenting. In the *Sorcerer II* project, for the first time Venter switched to open science practices, both for data sharing and the publishing of scientific papers. Craig Venter's public communication activities and strategies reveal the scientific cultures he interprets: that of an eighteenth-century 'savant' and nineteenth-century Victorian naturalist devoted to the exploration of new worlds, and that of a hacker, hero of an informational capitalism in which sharing is just another business model and independence from institutions means the escape from incumbents' control in order to open up new markets.

Emphasising his independence from both academy and industry, but building strong alliances with both spheres and with the media, Craig Venter sailed the seas of contemporary biocapitalism and media, interpreting a specific typology of the relationship between science and society, enterprises and universities. Indeed, one of the phenomena we are witnessing in contemporary science is the birth of new hybrid figures such as scientist–politicians and scientist–entrepreneurs who want to be a part of the academic community as well as other social groups. The approach to intellectual property, secrecy and, more generally, to information sharing, has become mixed and complex and the boundary between an academic biology devoted to sharing and disinterestedness and a corporate science based on secrecy and profit-making can blur into more complex innovation models. Furthermore, recent developments such as the rise of a hi-tech gift economy, an economy that is able to extract value from freely circulating information, has changed the way business is conducted in the informational economy. The border between academic and corporate biology has become so thin, with respect to sharing and intellectual property rights practices, that crossing it is no longer a cultural adventure.

In the *Sorcerer II* case Craig Venter – the bad boy of science, who was used as a symbol of science commodification and aggressive intellectual property rights policies – pushes those boundaries, as he seems to have a transformative agency that structures today's bioeconomies based on open information sharing. Venter also represents a type of science in which data management policies, sharing and communication practices internal to the scientific community are strictly intertwined, or even inseparable, from public communication dynamics. In the *Sorcerer II* case, Venter shifted from closed to open approaches to data and information sharing and, as I suggest, he represented only the tip of the iceberg of a new model of science–society interaction, rooted in the spheres of marketing, commercialisation and communication.

In his voyage there are important issues of contemporary science at stake, such as secrecy, access, exploration and the future. Actors such as Nobel laureates, mammoth web companies, wealthy venture capitalists, television producers and millions of genes take the stage in this story. Finally, characters such as the hacker, the entrepreneur and the Victorian natural scientist are depicted in the media portraits of Craig Venter and his boat, while open source and open access practices contaminate his strategies.

Several scientists make strong use of the media, are entrepreneurs and invest energy in developing links with politics and industry. But few decide to live in as many territories as Craig Venter did, or are able to build public communication practices in which they assemble rhetorical blocks coming both from classical elements and typical contemporary leitmotifs. Through the analysis of the relation between the *Sorcerer II* and the media, a specific image of 'scientist in public' represented by Venter emerges, particularly in relation to his decision to switch to open science tools, such as an open access database and open access journals.

The public communication strategies used by Venter are powerful instruments in the debate on the limitations, opportunities and interests to be favoured in today's biology. Based on data coming from discourse analysis, I draw attention to the media production linked to the expedition of *Sorcerer II*. The event received wide media coverage, and I collected the major international communication production that dealt with the *Sorcerer II* since the beginning of its voyage in spring 2003 up to the publication of the first set of results in spring 2007: my sources include J. Craig Venter Institute (JCVI) websites and press releases, but also press articles, TV programs, documentaries, interviews, scientific publications and books.

In the 1990s Venter participated in the race to sequence the human genome and has since focused on synthetic biology and personal genomics. He is considered one of the most influential scientists in the world and represents the emergence of science entrepreneurship and the privatisation of biomedical science. While working at the US National Institutes of Health (NIH), Venter contributed to the development of a technique for identifying DNA fragments called expressed sequence tags (ESTs). The NIH initially were oriented towards patenting ESTs, a decision Venter backed, but later it decided to withdraw the patent applications.

After leaving the NIH in 1991, Venter founded The Institute for Genomic Research (TIGR) and Celera Genomics, the private firm that challenged the publicly funded Human Genome Project and sequenced the human genome in 2000. Celera's shotgun sequencing techniques were in fact faster (but less accurate) than previous techniques. The company planned to create a genomic database to which scientists could subscribe for a fee. In 2000, Venter and Francis Collins, head of the HGP, publicly announced the mapping of the human genome, along with former US President Bill Clinton and UK Prime Minister Tony Blair. Venter became a symbol of a new kind of scientist–entrepreneur, portrayed in a famous photograph published on the cover of *Time Magazine* in which he wears a lab coat *and* a dress suit. After being fired by Celera in 2002, as the company's business model did not prove efficient, he founded the J. Craig Venter Institute and Synthetic Genomics, Inc., which corresponded with a new phase in his career in which Venter switched to more applied research, such as synthetic and personal genomics, and to a new socio-economic configuration.

The *Sorcerer II* is part of his focus on synthetic biology and research on biofuel. A 95-foot sloop, designed to be a sports craft and turned into a research vessel, the *Sorcerer II* was operated by the J. Craig Venter Institute in the Global Ocean Sampling Expedition, a circumnavigation of the Earth carried out to collect and sequence the genomes of marine microbial organisms. The ship, also funded by the Moore Foundation, the US Department of Energy and the *Discovery Channel*, sailed for thousands of miles stopping periodically to collect microbial material from the oceans' waters. After a brief expedition into the Sargasso Sea in spring 2003, the main journey of the *Sorcerer II* set out officially from Halifax, Nova Scotia, in August 2003, wending its way into the Gulf of Mexico, on to the Galapagos Islands, past Australia and to South Africa. The vessel returned to New England in January 2006, after sailing for 17 months.

The samples collected were sent to the Venter Institute in Rockville Maryland for sequencing. With its 6.5 million genetic sequences analysed and 6.3 billion base pairs catalogued, the expedition created the widest metagenomic database in the world, called CAMERA,[2] and gave birth to a publication in *Science* and a special issue of *PLoS Biology*. Metagenomics is the study of genetic material collected from environmental samples and not from individual organisms or cultivated clonal microbial populations. Metagenomics' goal is 'to produce a profile of diversity in a natural sample' without assigning DNA sequences to a peculiar organism, but rather by classifying them according to the characteristics of the environment in which they are collected.[3] The scientific goal of the *Sorcerer II* mission was to collect and to catalogue an unprecedented quantity of 'genes' expressed in different environmental conditions in order to use them in synthetic biology projects.

The *Sorcerer II* expedition was accompanied by a great effort of communication to the general public through different types of mass media. I have identified public communication methods which are common in most parts of the contemporary research projects, such as press conferences and press releases, but also direct interaction with the general media such as the case of the documentary shot by the *Discovery Channel* on board the *Sorcerer II* (Conover 2005). The website for the expedition contained a tracker allowing users to follow the route of the vessel, informing them on its real-time position. James Shreeve, a *Wired* journalist and biographer of Craig Venter (Shreeve 2004a), went on board the *Sorcerer II* to write an article (Shreeve 2004b) published as a cover story in the August 2004 issue of *Wired*. The JCVI (2006) presented itself as an institution able to leave the ivory tower to appeal to the citizens, by stating that it devoted itself not only to the advancement of the science of genomics but also to 'the communication of those results to the scientific community, the public, and policymakers', putting research and public communication on an equal footing.

BEYOND DARWIN

The *Sorcerer II* mission is placed explicitly in a long tradition of scientific research voyages, which include the expedition of the *Beagle* of Charles Darwin and of the *Challenger* (Gross 2007; JCVI 2004b), an oceanographic expedition that circumnavigated the globe between 1872 and 1876, stopping every 200 miles to examine the marine waters and search for unknown organisms, precisely as the *Sorcerer II* did. Venter himself stated, when talking about the *Sorcerer II* in his autobiography, that 'I found myself in a new yacht, sailing new seas, and seizing new scientific opportunities' (Venter 2007, p. 331).

Thus, one of the images of science put forward by the *Sorcerer II* is the one of the 'savant' explorers, scientists who carries out research away from laboratories and academia. Their enterprises take place within nature in an effort to discover the mightiness and the spectacular features of the universe, which coincides with an exploration of the world and the shift in the frontiers of human knowledge. Their dedication to research is all-encompassing, and their groundbreaking goals are not only economic but scientific too. The participation of the *Discovery Channel* fell within its 'Discovery Quest' programme, an initiative to fund a 'new generation of scientific discoveries', as the website of the TV channel maintains. It is about funding 'researchers and explorers' (and in the case of Venter the two figures overlap) at the forefront. Their feats should be told so that they can capture the 'genius, obstacles and happiness' of moments of revelation so strong that they can 'change science' (*Discovery Channel* 2005). The voyage is not only one through nature's secrets, but also a personal adventure, a voyage of discovery *and* 'self-discovery' that changes human understanding of the world as well as the scientist's life and understanding of himself (Venter 2007, p. 345).

On 4 March 2004, the JCVI held a press conference to present the study published in the *Science* issue of that week, describing the first set of data on the samples collected in the Sargasso Sea. The day after, it was in newspapers all over the world. During the press conference Venter announced that at that very moment his *Sorcerer II*, converted into a research vessel, was at the Galapagos islands, spurring journalists to underline the link between his voyage and Darwin's one. The *Wired* headline on the cover explicitly mentioned the most important work by Charles Darwin: 'Craig Venter's epic voyage to redefine the origin of the species' (Shreeve 2004b). Also *Science* and *PLoS Biology* highlighted the similarities between the voyage of the *Sorcerer II* and the *Beagle*. One of the images published by *PLoS Biology* shows Craig Venter at the Galapagos, posing next to the Estación Científica Charles Darwin. The exploration was associated with the discovery of unknown worlds and the achievement of wonderful scientific objectives:

> there was obviously an unknown and unseen world in the oceans that could be vital to better understanding diversity on the planet, as well as potentially solving some of the planet's growing environmental issues, such as climate change. (Shreeve 2004b)

Likewise, all the narrations on the Global Ocean Sampling Expedition underlined the comparison with Darwin's voyage, as demonstrated by this excerpt from *Wired*: 'He wants to play Darwin and collect the DNA of everything on the planet'. In the documentary produced by the *Discovery Channel* (Conover 2005) the image of the explorer of new worlds also makes an appearance. Craig Venter is examining a map before exploring a tropical island, with the ocean at his back. Equipped as a scuba diver, he plunges into the waters of the Cocos Island while the voiceover says: 'Strange things from deep within the Earth are happening ... and Craig Venter is here to investigate'. To illustrate the images of the website of the Global Ocean Sampling Expedition there is a quotation from Khalil Gilbran: 'In one drop of water are found all the secrets of all the oceans'.

However, although the concept of explorer embodied by Venter may seem more imaginative than real life, his scientific objectives are focused on the most urgent issues of our time: 'Craig Venter is starting to wonder if the food we eat and the air we breathe might not come from the place we think'; and he has embarked on 'a global voyage of discovery that might impact you and your neighbourhood's fueling station'. This is stated in the *Discovery Channel* documentary (Conover 2005), while the images go from the ocean to a Shell gas station where Venter arrives driving a hydrogen-fuelled car, to fill up the tank with clean and free energy. 'Future engineered species could be the source of food, hopefully a source of energy, environmental remediation and perhaps replacing the petrochemical industry' (Venter 2005). Indeed, bacteria 'are the dark matter of life. They may also hold the key to generating a near-infinite amount of energy, developing powerful pharmaceuticals, and cleaning up the ecological messes our species has made' (Shreeve 2004b).

Yet Venter does not limit himself to using and underlining the analogy with Charles Darwin – he wants to go *beyond* Darwin, thanks to the technical instruments he has at his disposal and to his special view of the natural world: 'We will be able to extrapolate about all life from this survey. ... This will put everything Darwin missed into context'. The enterprise undertaken by Venter, indeed, has all the instruments to trace 'all life on Earth. And his journey is just begun' (Shreeve 2004b). The *Sorcerer II* has found 'more species in one sample area than Challenger found in its four voyage around the entire planet' (Conover 2005). So, while Venter plunges into the waters of Galapagos and approaches an iguana, the voiceover says: 'now Craig Venter visits this ecosystem swimming with Darwin's subjects and collecting life invisible to the instruments of the 1830s' (Conover 2005). If Darwin's work drove a change in the way we see the world, Venter is hoping the marine data he is collecting will do the same in years to come, as his technical ability and technological tools are far broader than the ones Darwin had. Venter's adventure is meant to change the world.

CRACKING THE OCEAN CODE

The images offered by Venter are rich in references to his role as information scientist, another type of explorer of new worlds. Several discourses about the use of information and data are related to hacking and open source software. Venter does not refer specifically to hacking or free software: I do not want to use the term hacker as a native category but rather as a heuristic device in order to understand the typology of open science politics Venter represents.

In this sense, he embodies different features of hacker cultures. First of all, his insistence on informational metaphors that go beyond the metaphor of DNA as a code, to state explicitly the direct relationship between genomes and *software code*. Thus, he refers to genomes using computer-related metaphors: 'this is actually just a microorganism. ... We need to know his operating system' (Venter 2005). His objective is to 'create the Mother of all databases' (Shreeve 2004b), because 'genomes are like software code. Like code, genomes can be mapped' and recorded in a disk: the passage from life to disk means that genomes become 'digital code ready for computer processing' (Conover 2005). Life is genetic information, and the scientist managing to unveil its code using the IT of contemporary biotechnologies will be able to grasp its secrets and to exploit it to the benefit of all humankind.

Beyond Venter's case, over the last few years the emergence of synthetic biology research and massive genome sequencing has brought back the use of extreme informational approaches to the genetic code. The artificial synthesis of genetic material is based on the production of digital sequences. Synthetic biology projects often aim at standardising biological material: one of the most famous examples is the Registry of Standard Biological Parts, a collection of standardised biological parts that 'can be mixed and matched to build synthetic biology devices and systems'.[4]

Furthermore, the genetic code has been used to store information not related to life. In 2012 Harvard biologist George Church encoded the text of his book in DNA, that was then read and copied 70 billion times (Church et al. 2012). This heavy reliance on DNA's informational characteristics matches the hacker's role as a discoverer of codes, of secrets guarded by coded languages that may turn out to be useful, wonderful and surprising. For hackers, cracking codes and solving problems with beautiful hacks is a goal per se (Levy 2010a). In the logo used by *PLoS Biology*, the *Sorcerer II* sails a sea made of A, T, C and G, the initial letters of the four nucleobases that make up the DNA code, and we should 'join him in his attempt to change our planet's future by cracking the ocean code' (Conover 2005).

In IT jargon 'to crack' means to unveil an encrypted code, a metaphor already found in other studies on the public images of biotechnologies (see Davies 2002). Cracking is what hackers do when they violate the access to a system. The ability of hackers is based on their skills to manage and manipulate information. Even the *Sorcerer II* is trying to crack a code that must be unveiled, also without knowing its immediate use. A few years later, Venter was engaged in another research project: the creation of an artificial microbial genome. Venter used his ability to synthesise the genome to launch a hacker challenge to other biologists. Anyone who cracks the code has been invited to visit and send an email to a web address whose URL is written into the DNA as part of the genetic sequence Venter synthesised (JCVI 2010).

As in the case of several founding myths of the hackers' world, there is no need to find an application for decrypted codes. Venter's claims are along the lines of 'We found 20,000 new proteins that metabolise hydrogen in one way or another. 20,000!' or 'We're just trying to figure out who fucking lives out there' (Conover 2005). The genetic code is a source of power in itself. Venter says: 'If I could boost our understanding of the diversity of life by a couple orders of magnitude and be the first person to synthesize life? Yeah, I'd be happy, for a while' (Shreeve 2004b).

Furthermore, in Venter's voyage information is depicted as a goal in itself, an adventure experience, and stopping people trying to improve its understanding or acting directly on its mechanism implies a dictatorship. Besides highlighting the importance of 'bare' information, in the narration on the *Sorcerer II* the taste for discovery is mixed with the pleasure of life: while to a hacker curiosity and freedom can be crucial driving forces, fun and hedonism are also part of the justifications that characterise the new ethos of capitalism based on flexibility, creativity and freedom from bureaucratic command (Boltanski and Chiappello 2005). Their desire for knowledge and self-management in their work makes amusement an important component of hackers' activities, whereas to their eyes bureaucracy and institutions acquire a negative image.

When some critics remarked that he should have used a proper and real research vessel and not his pleasure sailing boat, which 'looks and feels pretty much like a luxury yacht' (Shreeve 2004b), Venter replied that he wanted to 'combine work with pleasure', sarcastically underlining that he 'will be joining the vessel very soon to head to French Polynesia. It's tough duty'.[5] In fact, the boat was named after the *Sorcerer*, Venter's previous ship, with which he had won the transatlantic Atlantic Challenge Cup, and which he had sold just before engaging in the Celera challenge to the HGP. The headline for the article on the *Sorcerer* published in the *Economist* was: 'What Dr. Venter did on his holidays' (*Economist* 2007). After all, the expedition left from Halifax in Nova Scotia because Venter 'had never sailed that far north and wanted to see what it was like' (Shreeve 2004b).

Wired, the magazine that sent a journalist on board the *Sorcerer II*, contributes to this image of a scientist. *Wired* was targeted by Barbrook and Cameron in their 'Californian ideology' (1996) as the herald of the Silicon Valley model of the relationship between research, technology, society and capitalism, a model that glorifies the garages in which young hackers develop their digital creativity and the headquarters of the venture capitalists, ready to pour millions of dollars into innovative projects with a high social relevance. Venter also has had direct contacts with the IT innovation companies. In *Google story*, David Vise and Mark Malseed describe the meeting between Venter and the two founders of Google, Larry Page and Sergey Brin. Today, according to Venter's vision, the real challenge of biology is to organise and analyse the huge quantities of data contained in the genetic databases, and 'Google's mathematicians, scientists, technologists, and computing power had the potential to vault his research forward' (Vise and Malseed 2006, 285). In fact, 23andMe, one of the most important personal genomics companies in the world, has strong links with Google and was founded after the meeting described above (see Chapter 2).

CAPTAIN HOOK IS GIVING THE TREASURE AWAY

However, hackers also have a business model and a reference market. Craig Venter is well-known for having adopted secrecy and privatisation policies for genetic data, yet in the *Sorcerer II* case he switched to a very different information management model. At the beginning of his entrepreneurial career with Celera Genomics he challenged the rules of academic science, forcing *Science* to change its publication standards and allow him to publish the study on human genome sequencing without making all of the genetic data public. Venter contributed then to the redefinition of the very concept of scholarly publishing. This is not, as Yurij Castelfranchi explains,

> because of the (old and given) fact that the private sector (with its patents, industrial secrecy etc.) could delay or stop the disclosure of scientific data, but more so, because of the (less known and rather new) fact that the historically important feat of certain scientists could be publicly recognised (and that they could enjoy the ensuing academic prestige), even though all

the related data were not freely available to their colleagues. (Castelfranchi 2004, p. 2)

In fact, Venter decided to publish in *Science* and the publication gave him an unheard of opportunity by not forcing him to deposit the Celera Genomics' human genome data in GenBank. Instead of following a typically corporate policy that often means keeping data secret without publishing any study until a patent is obtained, or until publication does not harm the profitability of a discovery, he wanted both scientific recognition *and* industrial secret.

Venter, in sum, managed to convince one of the most prestigious scientific journals in the world to change its publication policies. This was initially announced by the journal as a very special exception. Yet over the following years this behaviour was repeated for other papers. It therefore comes as no surprise that during the subsequent *Sorcerer II* voyage several people were still conscious that 'Darth Venter' had tried to privatise the human genome, 'allowing access to the code only to the deep pockets who could afford it' (Shreeve 2004b).

But approaches based on industrial secrecy, intellectual property rights and providing services based on open access data are becoming increasingly important modes of making money from biological information. The three are actually crucial in life science today (as well as in software and information technologies), but their respective weight changes and oscillates dramatically with time and in different areas. Fear of anticommons effects, national and international regulation, market demands, and public opinion are among several factors that influence the choices of what, when and how to appropriate knowledge in life sciences (Mills and Tereskerz 2007).[6]

Indeed, if Venter was at the centre of a furious polemics while working at NIH – NIH having filed two patent applications in 1991 claiming 4,000 of fragments of human DNA (ESTs), with Venter as the inventor – today he also insists on aggressive patenting tactics. In 2007, while publishing the Global Ocean Sampling Expedition papers in Public Library of Science and its data in the open access database CAMERA, the JCVI (2007) also filed a patent application for a 'minimal bacterial genome'. The patent application was so broad that a group of opponents of life patenting, the ETC group (the Action Group on Erosion, Technology and Concentration), compared Venter with the software corporation famously averse to openness, Microsoft:

We believe these monopoly claims signal the start of a high-stakes commercial race to synthesize and privatize synthetic life forms. And Venter's company is positioning itself to become the 'Microbesoft' of synthetic biology. Before these claims go forward, society must consider their far-reaching social, ethical and environmental impacts, and have an informed debate about whether they are socially acceptable or desirable. (ETC group 2004)

Of course, beyond the case of the *Sorcerer II*, the issue of making money from information remains at the heart of the scene. Yet Venter has chosen a different stance, deciding to release all data in the public domain and to publish the main results in *PLoS Biology*, a leading journal in the open access movement (Pottage 2006; Rai and Boyle 2007). Venter has underlined many times that he intends to produce data that anyone can freely explore 'from their desktop' and make it 'publicly available to researchers worldwide. ... No patents or other intellectual property rights will be sought by the Institute on genomic DNA sequence data' (JCVI 2004a).

Nonetheless, accusations of biopiracy came almost immediately (see Pottage 2006), when Ecuador and French Polynesia, whose territorial waters were crossed by the *Sorcerer II*, opposed the sampling because they feared it was an attempt to exploit their genetic resources. An agreement was reached between the Polynesian authorities and Venter himself after lengthy negotiations with the French government. In the meantime, Venter was criticised with the document *Playing God in the Galapagos* by the ETC group (2004) and was also nominated for the Greediest Biopirate award by the American Coalition Against Biopiracy (2006), winning the Captain Hook Awards 2006 'for undertaking, with flagrant disregard for national sovereignty over biodiversity, a US-funded global biopiracy expedition' on 'his pirate ship'. And yet, Venter presented himself as a defender of open access to scientific data, and rejected the biopiracy accusations:

he's doing everything he can to convince the world that he has no commercial motive: *Here, take it all, I ask for nothing in return.* His generosity has actually exacerbated his political problems. ... In return for access to their waters, governments expect a piece of the action. But if – like Venter – you are giving everything away, you don't have any benefits to share. 'The irony is just too great,' he says. 'I'm getting attacked for putting data in the public domain'. (Shreeve 2004b)

Furthermore, science as a whole was presented as being under attack, as well as progress towards new frontiers of knowledge. In Venter's discourse, anti-scientific obscurantism occurs when a scientist is forced to 'navigate the complex legal territory ... "If Darwin were alive today trying to do his experiments, he would not have been allowed to,"' says Venter. The comparison with Darwin's voyage is thus publicly used also to reject the accusations about the expedition: 'If it's in the Darwin school of biopiracy, then fine' (Nicholls 2007). Here, the future is at stake: 'If you do not perceive the possibilities in this shift, if you say *no* instead of *yes*, you will be left in the past. There will be whole societies who end up serving mai tais on the beach because they don't understand this' (Shreeve 2004b).

The solution to the problem of the short-sightedness of governments and NGOs that want to defend their genomic and biodiversity resources from the passage of the *Sorcerer II* lies in Venter's capacity to connect to the world of politics and, when needed, to mobilise it: 'He didn't sound too worried. He had already enlisted the French ambassador to the US to lobby Paris on his behalf, and some top French scientists were writing letters of protest to the ministry' (Shreeve 2004b).

Therefore, Venter's narrative with respect to data sharing and intellectual property is different here from the one he shows in other strategic fights. This change is brought about by shifts that occurred in the meantime in the realm of genomics: the largest databases in the world are now open access, and private enterprises sell services linked to the management of raw data. The Californian company DNAnexus, founded as a Stanford University spin-off, provides a cloud-based data analysis service to research companies and universities, and its platform is designed to allow customers to easily share DNA sequence data. Sage Commons is a public repository for genomic data open to scientists and research institutions in order to 'facilitate cooperative compilation, comparison and evaluation of network models of disease'.[7] Sage Commons is composed by very large datasets and tools for biomedical research, and was funded and provided with datasets by pharmacological companies such as Merck, Pfizer and AstraZeneca. It is meant as a pre-competitive environment that individual companies can feed upon in order to develop and appropriate new goods and services.

The business model put forward by Venter is linked to a service economy. While in the *Sorcerer II* case he guarantees everyone access to his codes, he also sells his company's services and know-how. An open source model of capitalism often reappears in the narrations on the expedition of the *Sorcerer II* and is presented as a crucial instrument for innovation. Yet while Venter's sharing practices embodied in the CAMERA database reflect the emergence of for-profit innovation models based on open science tools, they are also tools to form alliances with a very diverse set of funding sources, media and scientific institutions.

IN THE OPEN OCEAN

None of the images evoked by Venter are innovative, yet the recombination Venter makes of them is innovative. So, by making use of different strata and levels in the complex repertoire of popular imagery on scientists, and several strong metaphors, rhetorical tools and discursive leitmotivs, Venter manages to embody multiple figures and stereotypes of the scientist: the 'savant' explorer of the eighteenth and nineteenth centuries and the hacker of the third millennium, the 'amateur', the curious searcher of the truth enacted during the construction of academic science, and the ambitious, proactive, individualist *Homo economicus* of the knowledge society. Thus, the biohacker Craig Venter fully represents the neoliberal side of open science – one dominated by creative destruction and Joseph Schumpeter. The transformation of his data sharing policies is part of his shift towards a more open entrepreneurial environment in which different actors and different forms of intellectual property management coexist.

Yet Venter's open science is far from both neoliberal claims for the free flow of information and countercultural statements such as 'information wants to be free'. The analysis of the media narrations on his work shows a self-portrait in which an ambitious, brave, restless bioentrepreneur manages to get free from the institutional and bureaucratic constraints typical of twentieth-century science, bypassing what is considered to be the 'classical' figure of a modern scientist: linked to academia, disinterested, far from mingling with society and the market, belonging to a global scientific community made up of peers. To put it another way, he does not belong explicitly to industrial research and development, yet he is external to the stronghold of 'high' science: 'My greatest success is that I managed to get hated by both worlds', Venter says (Shreeve 2004b).

Big Bio scientists are part of a priesthood he wants to defeat with the help of information technologies:

> instead of having a few elitist scientists doing this and dictating to the world what it means, with Google it would be creating several million scientists. Google has empowered individuals to do searches and get information and have things in seconds at their fingertips. (Vise and Malseed 2006, p. 286)

Yet these claims, focused on anti-bureaucracy and openness, are always linked to the ability to raise and manage money from several different classes of funder, such as private companies (Google), governmental agencies (the US Department of Energy), non-profit actors (the Moore Foundation), and even the media (*Discovery Channel*). In the history of computers, hackers have gone as far as to get 'dirty money' from DARPA, the US defence research agency, and the CIA (Altman 2012; Levy 2010a), and Venter's mixed and complex funding scheme is not that different after all from the ones that sustained the birth of the computer industry.

In Venter's history, most of the images that include autonomy, entrepreneurship and individualism are hardly new: his withdrawal from the NIH in 1991 and from Celera Genomics in 2004 have allowed him to say he does 'any kind of science' he wants 'without obligation to an academic review panel or a corporate bottom line' (Shreeve 2004b). Venter's science is also embodied by the status of the institutions led by him: on the one hand, the J. Craig Venter Institute, a non-profit organisation, on the other hand Synthetic Genomics Inc., a company whose aim is to market (and, eventually, patent) the results of research projects on synthetic life.

His economic purposes, however, are always made explicit. As *Wired* has reported, being accused of pursuing fame and fortune, Venter 'cheerfully agreed'. Several scholars that study collaborative online peer production highlight how the information ecology that sustains it can include private corporations working for profit (Bauwens 2005; Benkler 2006). Companies can find several good reasons to contribute to a common pool of information resources released in the public domain or under open licenses. As we have seen in Chapter 2, over the last few years several private companies that rely on open science practices have emerged in the biomedical sector (see Hope 2008). These companies can participate in gift economies that rely on giving and data sharing as alternative business models. Yet, like social media websites that provide free services, they can also use openness as a marketing tool or as a way to increase financial capital (Arvidsson and Colleoni 2012). The *Sorcerer II* was able to integrate an open science economy based on sharing through open access databases and journals within a broader and more complex knowledge production system, thus contributing to the structuring of today's bioeconomy beyond the mere exploitation of gift economies.

In the narrations on this research project, having left academic science and industry aside, the *Sorcerer II* can finally sail the complex waters of the informational economy and of the new economic configuration of life sciences. In its voyage the ship has embarked, metaphorically or having them installed on board, on IT technologies to sequence and store data, biological machinery, journalists, cameramen, bioinformatics scientists, technicians, public research agencies, universities, start-ups, biologists, ambassadors, renowned scientists, non-profit foundations and private companies.[8] Contemporary biotechnologies require the creation of large and varied hybrid collective groups which make them multidisciplinary, connect them to private and public capitals and direct them towards the social needs expressed either by semi-public actors, such as foundations, or society in a broad sense (see, for example, Rabinow 1999 and Gibbons et al. 1994).

Of course, public communication is one of the tools by which these collectives negotiate their interactions. The *Sorcerer II* case does not represent a break in the norms regulating the production of scientific knowledge, but it interprets and drives changes going through it. Venter is excessive, and perhaps especially in the case of his vessel: he represents a science turned into a show, highly mediatised, barefaced as regards its objectives and capable of using sophisticated marketing instruments to discuss its work in the public arena to legitimise it and to give credit to its promises and results.

This chapter hardly represents a complete description of Venter's multidimensional and deep role in changing contemporary life sciences. Yet, though excessive and extraordinary, he is not a symptom of an illness in the relation between science and society, but rather an expression of its present physiology in a strategic area of science such as genomics. At the same time, the open science practices put in place by Venter do not represent a complete shift towards data openness. In 2007, the first high-quality diploid sequence of a human genome was published in the open access journal *PLoS Biology* while the data were put into the public domain: it was Craig Venter's. The JCVI is also providing researchers with JCVI Cloud

BioLinux, an open source tool for genome analysis to be performed through cloud computing platforms.

But this open science shift does not exhaust Venter's range of access and intellectual property rights policies. Over the last few years, the JCVI has been working on the production of a bacteria containing a synthesised genome, whose name is *Mycoplasma laboratorium* and is nicknamed 'Synthia'. Synthia was developed in order to produce a blueprint for a bacteria in which to insert the gene sequences needed for the production of biofuels or drugs (Gibson et al. 2010). Several patent applications for Synthia were filed by JCVI at both the US and the international level. Finally, while his data management and funding models are innovative, Venter is very careful in following academic publishing norms.

In studying other highly mediatised biotechnologists such as Kari Stefansson and James Watson, Michael Fortun (2001) has described the promises made by contemporary genomics as speculations on possible future scenarios. Miltos Liakopoulos has dealt with Big Biotechnological research projects such as human genome sequencing, identifying some recurrent frames into which the most frequent metaphors are grouped. In particular, he highlights the importance of the metaphors linked to the idea of 'progress', which present biotechnologies as a revolution that 'denotes a sudden break with the status quo and a fast rate of social change that, although dubious about the final effect' announces 'the violent change from the pre-existing order into a new, promising era' (2002, p. 10).

Brigitte Nerlich and Iina Hellsten (2004, p. 266), on the other hand, have defined the presence of metaphors linked to the human genome project as a 'treasure' or a 'landscape of opportunities' which should be explored: 'The metaphors of science as an adventurous journey, in which scientists venture forth onto a new "plain" ... with their trusted, but now seemingly complete, map in hand' seem to carry the greatest promises of the future of genomics.

It is possible to trace all these images of biotechnology in the public story of the *Sorcerer II*. Indeed, these are general and hegemonic commonplaces in the discourse of contemporary science. For Micheal Polanyi (1962):

the Republic of Science is a Society of Explorers. Such a society strives towards an unknown future, which it believes to be accessible and worth achieving. In the case of scientists, the explorers strive towards a hidden reality, for the sake of intellectual satisfaction.

And as Donna Haraway (1988) argues, Western science continues to be an important literary genre of exploration and travel. On the other side, as pointed out by Paul Rabinow (1999, p. 17) the argumentation that, with science and progress, our 'future is at stake', is crucial in contemporary narratives, especially in life sciences.

Venter seems able to feel the possibilities hidden in such metaphors, and to transform each leitmotiv in an epistemic tool, a powerful political argument or a marketing trick. During the years in which the voyage of the *Sorcerer II* took place, Venter was the promoter of other highly mediatised research projects,[9] acquiring credit among the general public as one of the world's most renowned scientists. Yet the scientific and media stage has seen the appearance of other biotechnologists using the same metaphors, exploiting the same images of science, keeping their balance on information disclosure and privatisation, and exploiting the media and the Internet in the same way.

Those scholars who have analysed the discourse practices of post-genomic biotechnologies have underlined the importance of these narrations. Michael Fortun (2001, p. 145) stated that the value of the new genomics companies are 'story stocks' dependent not only on genetic technologies but 'on that other set of technologies for simultaneously producing and evaluating anticipated, contingent futures: literary technologies'. Also the narrations on the *Sorcerer II* suggest a scientific, communicative and economic model, as well as a horizon to look at: the future.

Yet other scholars have addressed the *Sorcerer II* case from a different perspective. Alain Pottage (2006) analysed its effects through the process of bioprospecting, focusing on how new types of genetic collections are emerging in the age of bioinformatics and synthetic biology. Deterritorialising genes – that is, displacing them from their role in making living organisms – the *Sorcerer II* is a tool at work in the reconfiguration of current biocapitalism. Stefan Helmreich (2007) argues that Venter's ship is a means to virtualise oceanic genetic resources in order to create a new, empty territory for capitalist exploration: a new American frontier.

In my view, the *Sorcerer II* enterprise also shows that in some cases this deterritorialisation needs genetic information to circulate in open forms. In Venter, this need for a new open frontier mirrored deep transformations of his public ethos. His already known claims, related to autonomy from bureaucratic constraints and rebellion against academic dynamics and the modern science ethos, were remixed with hints of the traditions of both Victorian scientists and open source software practices. The practices enacted by Venter represent the neoliberal side of open science politics, a form of biocapitalism in which information sharing and circulation, both through open access databases and strong media relationships, lie at the centre of his mode of accumulation. While Venter may not be interested in enacting free and distributed cooperation, his open science shift has allowed him to have access to new forms of funding, as well as to new media alliances. Openness in the *Sorcerer II* case was never ambiguous: it was always related to individual and market freedom.

5
Just Another Rebel Scientist

He who guards against the lust and license of the pardon-preachers, let him be blessed!

Martin Luther, 1517

This is the story of a scientist, her challenge to scientific institutions and her role in the battle over transparency and access to biological information. It is also an example of the restoration of the twentieth-century open science ethos, one in which the re-emergence of the old ways of moralising about open science takes centre stage but is coupled with a willingness to change the way scientific institutions work.

Ilaria Capua is an Italian veterinary virologist who, in 2006, during the global avian influenza crisis, engaged in a clash with the World Health Organization (WHO) and its database access policies. Her actions, staged in the media and backed by an important community of researchers, eventually pushed the WHO to change its policies in favour of an open access model. Capua urged her colleagues to refuse WHO's policies, founded a new open access database under the umbrella of the Global Initiative on Sharing Avian Influenza Data (GISAID)[1] and became a well-known advocate of open science (and open biology in particular).

During the events that led to the creation of GISAID, Capua somehow came to personify a conflict among institutions, a conflict that took place in the media and whose resolution required the mobilisation of specific scientific cultures. Her story is an example of the role that a scientist's public ethos can assume in a context where the relation between researchers and scientific institutions is at stake, and in which public communication is the main site where the debate about open science takes place.

In this chapter, I try to show how the modern science ethos can still be used as a tool in the debate and negotiations about open science politics, even though it needs to be updated with the inclusion of elements coming from a different cultural tradition. Indeed, in the first instance Capua's cultural material is represented by the set of values inscribed in the Mertonian norms – the classic twentieth-century scientist's ethos. Yet during the events that gave birth to GISAID, Capua updated this old ethos through both a transformation and a definite continuation of it: a pre-existing culture that is dynamic and not static and whose changes show a peculiar facet of open science politics.

In simultaneously studying the narrations around Ilaria Capua and her practices, I seek to understand how the cultural toolkit at her disposal – the Mertonian science ethos – has been reshaped and adapted so as to give rise to new strategies of action. As I suggest above, during moments of crisis, such as Capua's clash with the WHO, actors must refer to pre-existing cultural models and revise them. In Capua's case, the occasion for reconfiguring the scientists' ethos was the need to restore openness by urging a big international institution to change its data sharing policies.

Scientists' agency in the establishing of open access policies has often proven to be effective: in 1996 several leading geneticists met in Bermuda in order to discuss data access policies and agreed on sharing human genome data online before publication in scientific journals. The Bermuda Principles shaped subsequent policy decision, in particular the Human Genome Project policies related to access to raw data. Yet Capua's case is peculiar, as *control* over information seems to be at stake more than mere sharing. The face of Big Bio that emerges from her discourses is one of a mammoth bureaucratic public institution challenged by a rebel biologist. There is the dream of a new type of open science enabled by the Internet and ICT tools at stake, one in which international cooperation can be enacted by online data sharing. Barriers to access, in Capua's story, put the brakes on collaborative science's possibilities and make researchers' efforts less productive.

The story of GISAID's creation is always referred to as a story of rebellion. Indeed, the scientist who rebels against authority is a common image throughout history and in the philosophy of science and biology (Dyson 2006; Harman and Dietrich 2008; Kuhn 1996). How would you start a rebellion if not with a public call to disobedience? That is what Capua did in questioning the WHO, eventually forcing it to change its policies on the access to avian flu data and establishing GISAID and its EpiFlu database, now one of the main global open access databases for flu viruses.

In this chapter I want to ask if Ilaria Capua is just another case of a rebel scientist belonging to that tradition, or whether there is something more in her public image and in her practices. In order to answer this question I turned to the media and communication production related to the birth and establishment of GISAID. Ilaria Capua had a major role in the events that preceded and accompanied GISAID's birth, which received wide international media coverage. Following an analysis of Ilaria Capua's practices and the history of GISAID's creation, I take a cultural perspective and use the hacker ethic as an analytical tool to interpret some core aspects of the open science culture expressed by Capua. Although she does not describe herself as a hacker, I believe that comparing her public virtues to those of the hacker ethic is useful in order to understand whether, and in what ways, she interprets a specific and emergent typology of a scientist, as well as a specific type of open science. My research material is composed of a review of four years of the major Italian and international communication production that dealt with Capua and GISAID, beginning with the letter she sent to her colleagues in February 2006 that was reported by the journal *Science*, and ending in 2009. My sources include websites, press articles, radio programmes, press releases, emails, conference talks, interviews, scientific publications and books, several of them directly written by Capua. I also conducted an in-depth interview with Capua.

Ilaria Capua works at the Istituto Zooprofilattico Sperimentale delle Venezie (IZS), an agency within the Italian public health system based in Padua. Her work as a virologist was already well-known nationally and internationally thanks to her role in the vaccine field and to her research activity on avian viruses. In 2001 she developed the DIVA vaccination strategy against avian flu.[2]

Before getting involved in the WHO case, she authored or coauthored dozens of scientific papers, mainly in veterinary science journals such as *Avian Pathology*, *The Veterinary Record* and *Avian Diseases*. In 2005, after having already been appointed to several other national and international positions she was nominated Chair of the Scientific Committee of OFFLU,[3] an FAO/OIE[4] agency established to fight avian influenza. Yet Capua became famous globally during the public health avian flu crisis between 2005 and 2006 and the media coverage that surrounded it.

In January 2005, an outbreak of avian influenza caused by a highly pathogenic mutated strain of the H5N1 virus hit Vietnam. Over the following months, the outbreak spread to China and other Asian countries. In 2006, the virus was detected in Africa and finally in Europe. At the time, deaths by H5N1 infections caused by bird to human transmission of the virus were beginning to appear, and a pandemic evolution was considered possible. Media and political debate were focused on this scenario.

Capua's story begins with a letter she sent in the middle of the crisis. Her lab was a reference centre for the FAO and OIE, and in January 2006 it found itself needing to deposit data that related to the sequencing of some H5N1 strains (the avian flu virus). One of them was from Nigeria (the very first diagnosis occurred in Africa) and the other from Italy. On 16 February 2006 Capua sent an email to 50 colleagues urging them to refuse the WHO policy – until then the WHO published genetic sequences of the H5N1 virus in a database with access restricted to only a few research groups working with the organisation. Other important institutions had by then already established coherent and broad data sharing policies: for example the NIH,[5] the Human Genome Project[6] and the US National Human Genome Research Institute, while private institutions such as the Wellcome Trust were also adopting open data policies.

However, for several international agencies such as the WHO and FAO, no general sharing agreement was in place. According to Capua, virologists and geneticists working on H5N1 should instead deposit their data in the public and open access database GenBank, rather than on the WHO database. GenBank collects all publicly available nucleotide sequences and is produced and maintained by the National Center for Biotechnology Information (NCBI), an initiative of the NIH in the United States. GenBank receives sequences produced in laboratories throughout the world and is the most important database for research in the life sciences.

So Capua decided to put her own data in GenBank on that very day. On 3 March the news about Capua's stance was published by *Science*, the first magazine to write about the WHO–H5N1 affair (Enserink, 2006a; see also Capua et al. 2006). A few months prior to this, in September 2005, a similar critique involved the US Centers for Disease Control and Prevention (CDC), which was being accused of not fully sharing its avian influenza data (Butler 2005). But at that stage the debate had not reached the general media. This time, a few days after the publication of her letter, the clash between Capua and the WHO was staged in public, not just through media directed at a professional public or at a community such as the mailing list ProMed, which was one of the first arenas in which the debate took place.[7]

Over the course of the following weeks and months, the debate not only involved major scientific journals such as *Science* and

The Lancet, but also major US opinion-leading newspapers such as *The New York Times*, *The Washington Post* and *The Wall Street Journal*, and magazines such as *Scientific American* and *Seed*. In Italy almost all national newspapers and magazines covered the story (*Il Messaggero*, *Il Corriere della Sera*, *La Repubblica*) together with several magazines such as *Le Scienze* and *Wired Italia*. In addition, *Nature* openly endorsed Capua's decision: 'Three cheers for Ilaria Capua' (Anonymous 2006b, p. 255).

The heads of the WHO and Capua's colleagues were therefore forced to enter the debate and position themselves. On 30 March, the first group of colleagues openly rallied in support of Capua with a letter to *Nature* (Salzberg et al. 2006). In June, following a second letter to *Nature*, US policymakers began to ask for a mandatory open access policy for H5N1 data, similar to databases such as GenBank. In August it was Indonesia's turn: the government of a country heavily affected by the avian flu virus removed all restrictions on access to its data.

Eventually, on 31 August 2006, Ilaria Capua and Peter Bogner, a strategic advisor who had joined her in her effort to build a new database, together with important scientists from the CDC and NIH, announced the creation of GISAID, 'a global consortium ... that would foster international sharing of avian influenza isolates and data' (Bogner et al. 2006, p. 981). Their letter, published in *Nature*, was cosigned by 70 scientists and health officials, including six Nobel laureates. GISAID proposed that geneticists, virologists, veterinarians and epidemiologists would agree to share their data by depositing them as soon as possible in a major open access database (such as GenBank). After the publication of the letter, Ilaria Capua was honoured with a profile in *Science* (Enserink 2006b).

Yet her final victory came in January 2007 when the WHO finally adopted a resolution, completely changing its policies by asking member states to 'ensure the routine and timely sharing' of flu viruses.[8] In May 2008 GISAID opened its own open access database in collaboration with the Swiss Institute of Bioinformatics and with the backing of the Swiss and Indonesian governments, the OIE and other private partners such as The Bogner Organization. GISAID collected data relating to avian flu, and 2008 was the year in which Ilaria Capua was publicly consecrated as a star of global science, both for her role in flu virus research and as a famous open science advocate. She won a *Scientific American* SciAm 50 Award and was nominated as one of the 'revolutionary minds' of science by the magazine *Seed*. She was also nominated as the 'veterinarian of the year' in Italy and has maintained a high profile in the media by publishing books, giving television and radio interviews and penning editorials.

REVOLUTIONARY MIND

Almost every narration on GISAID's birth depicts Ilaria Capua as a rebel and a revolutionary. The recurring terms used to describe her story are refusal, rebellion, revolution, blame and challenge. On the other hand, the WHO and the countries that opposed open data sharing are labelled with expressions that refer to secrecy, to an 'old-boy network' and a self-elected circle which needs to be broken. Nothing new because, as I have argued in Chapter 3, the rebel, iconoclast, maverick and heretic biologist is a classic element of the narratives on modern science and biology, and often the iconoclast becomes a public icon.

The media reported Capua as rebelling against science's institutions, first of all against the WHO, but also rebelling against publishing and recognition mechanisms that characterise the work of the scientist and the functioning of modern science. She says 'no to science's book of etiquette' (Oriani 2006, p. 71) and she does that by 'slamming her hand on her desk' with her 'in-your-face opinions' (Enserink 2006b, p. 918). Her image as a rebel scientist is saturated with statements such as 'I broke the moulds' (Capua 2009b). Also, media reports stress how she challenged institutions and fellow scientists – with her impassioned call she 'threw down the gauntlet to her colleagues' (Anonymous 2006a).

Rebellion, according to Capua's public image, is directed towards changing the way in which science institutions work and research is organised. In December 2008, *Seed* magazine included Capua in its special issue on 'revolutionary minds', nominating her as a 'game changer' of science. *Seed* highlights how Ilaria Capua is not 'willing to settle for the status quo' (Anonymous 2008, p. 82). Furthermore, her rebellion starts from an outsider's position. This is how *The New York Times* replays her mutiny: 'a lone Italian scientist is challenging the system by refusing to send her own data to the password-protected archive' (Anonymous 2006a).

In media accounts Capua occupies an underdog position. She is an outsider, a woman, an Italian and a veterinarian. She also positions herself as a pioneer who works behind the scenes and for whom 'the road is all uphill' (Coyaud 2007, p. 78). Although she does not work in a garage, as the hacker mythology imagines, her starting conditions and peripheral position are described as keeping her outside the inner circle of science. After all, the woman scientist as an underdog and rebel in a sexist world is another leitmotiv in the history of science. While phenomena such as the 'glass ceiling' are well-known for preventing women from occupying leading positions in science institutions, famous women scientists have been the object of debate about their rebellion against the status quo.[9]

Yet the narratives around Capua give it a peculiar spin. She is an underdog but some media coverage keeps highlighting how glamorous she is. Due to the acronym of a vaccine protocol she invented, they call her Influenza Diva (Enserink 2006b). Finally, Ilaria Capua is often irreverent and playful. She plays with information and does not look for formal recognition. In August 2006, after the launch of GISAID in the journal *Nature*, Capua commented 'I am so happy. I feel that maybe I should quit working and start arranging flowers' (Pearson 2006). When somebody told her she was going to be included as one of 2008's most important scientists, she answered: 'Really? That's weird, I was in the 2007 one'. Capua also affirmed 'I thought it was spam, since it seemed such an obvious thing for me to do ... Can you believe they give you a prize for doing something like that?' (Coyaud 2008a, p. 66). Sharing information is portrayed as a natural and fun thing to do, confirming her adherence to justifications related to the pleasure of working and innovating, as well as to the necessity of sharing information and knowledge outside institutional channels.

Capua takes a moral stance with her rebellion. She argues she is undertaking an 'ethical revolution' related to an unavoidable, individual choice: 'I find myself at a crossroads: to become one of the self-elected trustees of science, or to make our data available to the scientific community (Capua 2009b). Yet she does not rise up against an established knowledge system – hers is not a Kuhnian scientific revolution but rather a rebellion against institutions and their policies.

The rebel scientist can appear both within and outside the most important scientific institution: the university. Not all rebels must work outside the academy, yet often a rebel scientist has to make a break with institutions and the authority of her/his peers. In the case of hackers, enemies are often corporations, which are old, slow and hierarchical bureaucracies that are not transparent and open to public scrutiny and free competition. In most hackers' accounts bureaucrats are depicted as hiding behind arbitrary rules in order to avoid transparency and participation.

For the first generation of hackers the epitome of evil bureaucracies was a huge company called International Business Machines – or IBM – with big mainframe computers disdainfully called Hulking Giants (Levy 2010a). Corporations privatise creativity and slow down the innovation process. In the narrations linked to the birth of GISAID, as IBM and Microsoft are to hackers the WHO is to Ilaria Capua – an institution which shares many traits with big computer corporations. In this sense she is an antiestablishment character: while criticising the WHO and its data disclosure policies as well as the priorities of some countries, she highlighted how 'academic and national pride must not be allowed to slow potential crucial health research' (Anonymous 2006a), for 'results are usually either restricted by governments or kept private to an old-boy network of researchers linked to the WHO, the US Center for Disease Control and Prevention, and the FAO' (Anonymous 2006b, p. 266).

Capua asks her colleagues not to give in to the flattery of a 'fraternity' in which the priests are members of a 'select circle' that one cannot freely access and relate to (Enserink 2006a, p. 1224). Thus, Capua's publicly expressed distrust is not directed towards fellow researchers, but mainly towards scientific and political institutions and the rules a scientist has to follow to be part of them. Nonetheless, she is firmly part of public scientific institutions. She works for the Italian government, for international public agencies and is the head of several European projects. Her attack against the WHO is arranged with her institution's Director and with the General Director of the Italian Veterinary Agency.

Capua thinks she 'had the courage to do it, but also the freedom to do it ... I did not end up in Guantanamo'[10] exactly because she was an (Italian) underdog: scientists belonging to important institutions are not free to break the mould. Yet having unhinged a system, she claims to have made lots of enemies among people who work around the WHO or who belong to the political establishment – people who say 'yes, we remember you very well, because when that thing happened it blew up in our face'.[11]

However, on GISAID's website, along with the legal procedures and arrangements, scientists are asked to adhere to the scientific good manners that Capua shunned outright. Indeed, GISAID's platform 'is accessible to anyone who agrees to its basic premises of upholding a scientific etiquette',[12] resulting in collaboration, sharing and fair exploitation of the results. As a matter of fact, the GISAID case was a positive turning point in her career. During the following years Capua will be called upon to handle important tasks through major scientific institutions such as the CDC in Atlanta. With her increasing fame, there are even rumours that speak of her candidature as the Minister of Education, Universities and Research in Italy. Yet Capua, confirming her antiestablishment nature, makes known that she has 'tons of more important things to do' and even claimed to have turned down prestigious job offers by 'top brass' while she was a guest at an interministerial conference in Sharm el-Sheikh (Coyaud 2008b, p. 42).

Publishing in scientific journals subject to peer review is the primary and widely accepted validation tool for scientific knowledge. A scientist's authority – and his/her career – depends on formal peer recognition. But Capua introduces in her public discourse a critique of the peer recognition system, as she questions the publishing mechanisms of science. She accuses colleagues of being jealous and mean, of not putting the sequences they identify into the public domain 'unless they have already published results in a scientific journal, in fear of not having their work recognised or of losing their rights of economic exploitation' (Pistoi 2006, p. 31).

First and foremost, she claims they worry about personal success and about publishing in prestigious journals. They take advantage of the fact that avian flu 'makes audience'. According to her narration, that system pushes scientists to practice a sort of a publishing 'amongst friends' while they wait to submit to the prestigious publication that 'will bring the researcher glory and money, as well as hope and knowledge to the hoi polloi' (Oriani 2006). Nevertheless, the public ethic produced by Capua seems to drive her towards a different choice: 'what is more important? Another paper for Ilaria Capua's team or addressing a major health problem? Let's get our priorities straight' (Enserink 2006b). According to this choice, Capua gives priority to the avian flu problem over her own scientific career. She suggested that another research group had used the data she put in the public domain in a publication, causing her group damage from the point of view of academic acknowledgment and incentives. Capua thus 'has renounced the prestige of a distinguished international journal that would have given lustre to her career and has given priority to the speed of information ... regardless of rankings' (Calabrese 2006, p. 16).

The role of traditional scholarly publishing in peer reviewed journals is under attack here: publishing slows down data diffusion. Open science, in this sense, is a means to overcome common practices related to the way science's reward system is organised. The moral ground upon which open science should be built must be free from the selfishness instilled by the very organisation of scientific publishing. According to Capua, in the avian flu case publishing decisions were driven by scientists' personal needs or by academic, governmental or institutional interests. Indeed, the hacker pursues knowledge in a way that can be totally independent from hierarchies or institutional goals. The only acknowledgment he/she looks for comes from his/her results: to crack a code is a goal in itself, and to prove that your hack works is the only thing you need to validate your work. Hackers want to write good code, not to publish peer reviewed research papers, and they often value charismatic authority over formal and bureaucratic reward systems (O'Neil 2009).

Obviously several other biologists have criticised the peer review and scholarly publishing system and have tried to break or stress its rules. Sometimes rebels create their own independent publishing system, for example journals and other media subtracted from academic censure. In the 1960s, biologist Peter Mitchell chose to work in his own personal research institute, Glynn House, and to publish his work with Grey Books, a publishing house he founded himself (Harman and Dietrich 2008). More recently, Craig Venter forced the journal *Science* to change its policies for the publication of human genome sequencing data (Castelfranchi 2004, see Chapter 4).

Thus GISAID, Capua's answer to the canonic publishing system, proudly announces itself as a truly independent database, a database 'by scientists for scientists' where 'researchers like you have come together'.[13] Hacker communities (or recursive publics, in Kelty's broader definition, 2008) are based on the ability to create and maintain not only the content of platforms but also the legal and technical rules and infrastructures they need to spread and manage it among peers and to create the conditions for the very existence of the community.

Yet GISAID is also depicted as a return to an ideal, utopian scientists' community independent from external conditioning. Several scholars have highlighted that the publishing system for scientific journals does not satisfy all of the researchers' communication needs. In fact, in popular media they often find another important discussion arena (for a famous example see Lewenstein 1995). Capua herself seems to corroborate this hypothesis when she urges her colleagues to 'take out your sequences or get out of television news!' (Coyaud 2007, p. 78), or when she engages in an intense relationship with the media, giving dozens of interviews, writing editorials and also a book directed at students willing them to begin an academic career in veterinary medicine.

Within a few months she became an international 'media darling'. She describes her inclusion within *Seed* magazine's 'revolutionary minds' or *Scientific American*'s 50 best scientists as formal recognitions at an international level: 'two of the more prestigious prizes in the scientific world' (Capua 2009a), even though *Seed* and *Scientific American* are popular science magazines that do not have any scientific value according to any institutional parameter. Science institutions, with their recognition and incentive systems, are not the only world Capua lives within and gets legitimation from. Social recognition, even at the level of peer scientific recognition, is to be granted to scientists who decide to campaign for open biology. Indeed biomedicine is the scientific field where political and legal controversies over commodification and access to information have been the most harsh. This is why openness and freedom from bureaucratic constraints can be valued more than in other disciplines.

FREE THE DATA!

Beyond the freedom from bureaucratic constraints that characterise some scientific institutions, Capua's rebellion is directed towards the openness of avian flu data. While knowledge sharing is one of the main traits of the Mertonian ethos, Capua's claims are further developed towards a battle against secrecy. Indeed, the denunciation of secrecy can be ritualistic in modern science and is a deterrent to inquiring into the ethical problems arising from the choice between secrecy and openness (Bok 1982). Merton himself stated that secrecy was the antithesis to scientists' social need to share. Both trade and military secrecy were subject to open criticisms during the twentieth century, but this is not new. While referring to the seventeenth century and the scientific revolution, William Eamon highlights the political significance of the struggles against forbidden knowledge and secrecy:

> The rejection of secrecy in science was, in part, a reaction against what was perceived to be a closed, self-contained, and hierarchical system of knowledge, and against the official policies and institutions that maintained its exclusiveness. (1990, p. 356)

Transparency is also one of the most recurrent common features of hacker cultures. In the discourses that surrounded GISAID's birth data needed to be open, accessible and free for all, unlike the information which is kept secret or hidden by a plot or an institutional closed circle, let alone the password required to browse them. In the mid-twentieth century, Norbert Wiener's career culminated with an attack on the patent system and its relation to academic institutions as, he proclaimed, it 'impeded the flow of information in the great network that was society'.[14] Restriction to access was seen not as a mere cause of friction, but rather as 'deliberate jamming' (Johns 2009a, p. 426). At the MIT building 26, in the early 1960s, the first generation of hackers would find any possible way to sneak into locked rooms. Doors were just another obstacle between them and free information and hackers would copy master keys in order to be able to open any door, at night, when the building was left to them.

> The master key was a magic sword to wave away evil. Evil, of course, was a locked door. Even if no tools were behind the locked doors, the locks symbolized the power of bureaucracy, a power that would eventually be used to prevent full implementation of the Hacker Ethic. Bureaucracies were always threatened by people who wanted to know how things worked. Bureaucrats knew their survival depended on keeping people in ignorance, by using artificial means – like locks – to keep people under control. (Levy 2010a, p. 96)

Thirty years later, Richard Stallman would fight against MIT's attempts at keeping unauthorised users out of the system. Stallman famously started a 'password battle' during which he urged people to use an empty string as a password, in order to allow anybody to enter the systems and 'delay the fascist advances with every method' (Levy 2010a, pp. 439–41).

Contemporary examples of communication projects related to hacking, such as the global platform Wikileaks, have emerged as key players in battles over transparency. The relation to open science politics is more straightforward than it might seem. For example, John Sulston, the Nobel Prize winning biologist who sequenced the C. elegans worm and worked for the Human Genome Project, is a leading campaigner for open science and is against patenting genetic sequences. In 2012, Sulston was among the five guarantors required by the British court in order to grant bail to Julian Assange, Wikileaks' founder and spokesman (Hough and Bingham 2010).

Radical transparency is one of the main facets of hacker morality. In describing Richard Stallman's choices that lead to the creation of free software, Coleman and Golub talk about 'a liberal version of freedom that invoked the virtues of sharing and pedagogy' (2008, p. 261). Obviously, Capua spreads data and knowledge that are in her possession. Yet she also denounces and blames the machinery of secrecy that has to be broken: data is kept 'under wraps' and 'behind closed doors', and put into 'closed drawers'. Free the data! urges Le Scienze's headline (Pistoi 2006, p. 31). According to Ilaria Capua everyone with an interest should be able to freely access and browse all the data, and she 'prizes openness over secrecy, access over scarcity' Anonymous 2008, p. 83).

However, in Capua's discourse the reasons for action are civic, and the declared goal is the common good rather than any corporate or personal target. The currency of open science is not merely scientific reputation. In GISAID's case, one of the reasons for the WHO and some countries to oppose data diffusion is the need to prevent companies from appropriating or privatising data which could be useful to develop vaccines or tests against avian flu. Yet in the case of Capua the market sat in the background, without direct links with GISAID.

For example, private companies are able to develop and commercialise vaccines and drugs against the flu, as GISAID users acknowledge when they sign the database's Access Agreement.[15] Capua is not against patents per se, but rather is convinced that common sense urges institutions to share crucial data which are important for public health: 'they sometimes depict me as the Naomi Klein of science, but that is not true.'[16] The will to improve scientific productivity and speed is behind her choice, and not rebellion against private profits. Furthermore, the liberal ideal of free speech that informs free software production does not surface in this case. In the narratives about Capua openness is indeed valuable because it can be useful to defy avian flu in a time of emergency. Information collection is a goal that always needs to be coupled with its sharing, as it is something the world badly needs and the avian flu crisis is a moment in which information can be more valuable than ever. Given the magnitude of the threat, 'the current level of collection and sharing of data is inadequate' and openness seems to be directed broadly to all, to humankind, 'to the world as a whole' (Bogner et al. 2006, p. 981).

The world needs information, as confirmed by the fact that 'our data has already been downloaded more than a thousand times', and it needs 'real-time availability' (Cavadini 2006, p. 21). Information is good per se, even when it does not have any known goal, function or purpose or if the path one should follow in order to reach this goal or to fulfil its purposes is not clear. Indeed, the practices of several hackers are built around the need to crack a code, to unveil hidden information just for the sake of it. To Capua, an open, accessible database is a 'dream' that could help feed science, since there is 'hunger for information ... with my data another researcher could get to conclusions I can't even imagine' (Cavadini 2006, p. 21). To hoard genetic information in a database can be a goal per se, provided it is accessible, shareable information that can be used by other scientists: 'wait a minute, we're talking about a serious potential threat to human health. ... Not enough scientists have had the opportunity to look at this virus' (Anonymous 2008, p. 83).

The scientists who use GISAID must agree to 'share their sequence data, to analyse the findings jointly and to publish the results collaboratively'. The insistence on sharing and collaboration mirrors the license that researchers sign to get access to the database and upload or download data. Despite statements that highlight the willingness to make data accessible to 'all', GISAID data are accessible to all registered users, but not to others unless they have agreed to the same terms of use. Designed after a Creative Commons license and with the help of intellectual property expert consultants,[17] the database's access agreement allows scientists to 'reproduce, modify, disseminate' the data and author or publish results obtained from their analysis, as long as they give credit to the originating laboratory and GISAID. Yet they can not do the same with the EpiFlu database platform and software technology, which are proprietary and partially owned by third parties as well as protected by copyright. Contrary to open source data licensing, users *cannot* 'copy, reverse engineer, disseminate or disclose' any part.[18] Openness does not apply to every layer of the information environment: the content layer is open while the logical level, composed by standards and software, is subject to copyright (Benkler 2006).

NOT JUST ANOTHER REBEL SCIENTIST

During the global avian flu crisis that began in 2006, Ilaria Capua pushed the WHO to shift to open science policies for sharing influenza data. Her refusal of the WHO publishing policies and her role in the birth of GISAID and EpiFlu, the open access database for the sharing of avian flu virus data, have brought her fame, prizes and a reputation as an international open access advocate. While EpiFlu is but one of the dozens of open access biological databases that compose the data galaxy of contemporary life science research, what is interesting is the original alliance of scientists, policymakers, public institutions, foundations and private companies that was built by Capua, as well as her battle against the WHO.

Different scientific communities backed her efforts, most significantly veterinarians, epidemiologists and geneticists who joined her in this battle over data access and transparency. Yet Capua should not be considered *just another rebel scientist*, as her story cannot simply be assimilated with that of the traditional public image of the scientist as a pathbreaker and rebel against hierarchies of established knowledge. Rather, she represents a restoration of the old open science ethos related to Mertonian norms, and the re-emergence of an ancient and recurrent character in the history of science. Furthermore, this ethos has been subject to a cultural remix. Capua's public dimensions represent a reconfiguration of different cultural systems as she incorporated traditional scientific elements coming from hacking and free software cultures.

Capua exemplifies a remix between an old accepted culture embodied in a complete set of practices and norms and a more recent culture that we can see at work in several other fields of innovation. In this sense, public ethos is one of the possible cultural characteristics that individuals can mobilise when the need for a reconfiguration of different aspects belonging to one or more pre-existing cultures become more insistent. And this happens mainly in contexts of crisis and change such as the ones Capua went through. Indeed, as I argued in Chapter 1, cultural frameworks can both enable and constrain individual processes, providing actors with flexible yet resistant toolkits.

Contemporary scientists can still find cultural elements in the Mertonian science ethos that fit with their needs for the production of a successful public image because the influence of that culture still remains, even though the social dimension from which it was born does not exist any more. But they often need to be remixed with new and different ethical and cultural elements. Again, with regard to her role in the birth of GISAID, the individual elements of Capua's public image are not innovative, and yet the remix is innovative – this is the recombination that I have described.

Capua belongs to a longstanding tradition of scientists rebelling against established ideas and the upper echelon among their colleagues (see Harman and Dietrich 2008 or Dyson 2006). But she is not *only* another rebel scientist. Instead, she embodies a new and emerging figure of the scientist, one who uses some open source and open access tools more attuned to the current configuration of science and society relations. Yet at the same time she is not *only* an open access advocate. The call for the adherence to open science does not apply to every layer of the information environment she lives within, and it is always coupled with the rebuttal of bureaucratic control and claims of independence from both academic and corporate institutions.

In the narratives that depict her role in the birth of GISAID, several classic features recur, such as autonomy, independence and openness. But along with these, other characteristics emerge: the radical refusal of external interference and also of scientific institutions themselves; a component of hedonism; the insistence on bare information as a good per se, as long as it is shared and accessible; the importance of being an underdog; and an intense relationship with the media. Thus Capua attempts to shape scientific institutions, pushing them towards a transformation in the direction of a more open environment for the exchange of information – one in which the power of Big Bio bureaucracies is diminished in favour of a more horizontal model.

The WHO has been, in a sense, one of the last international institutions to resist the spreading of expressly open access policies. Finally, Capua rebels against the mechanisms of publishing and the peer review of scientific knowledge in the name of a type of cooperation that is enacted directly between scientists and not mediated by institutions. Her attempts to legitimise herself outside of some of the major institutions of science cause her to break their mechanisms. Her main interlocutors seemed to be an ideal scientific community formed by peers and communicating through the media.

With her insistence on openness and antiestablishment critique, and ultimately her role in restoring a lost ethos of science, Capua is once again making 'moral' the public image of a scientific field that badly needed a renewal after the anticommons crisis and the legal, political and societal clashes that came with it. Indeed, public communication is an important tool that scientists need to use in order to overcome social backlashes and to thrive in a demanding social environment (Bucchi 1998). The results she obtained in terms of shifts in institutional policies, institutional appointments and media prestige imply a peculiar ability to mobilise her public ethos as a tool for better positioning herself within scientific institutions that are in the process of transforming.

However, besides her personal results, Capua's case can be interpreted as a symptom of the emergence of a new open science culture in biology and genomics in which the restoration of a modern ethos devoted to a true science, based on open sharing, is of great importance. The case of Capua is particular – she is a scientist who works for public agencies and she is not strictly linked to private corporations. Hers is an example of an academic scientist only collaterally linked to corporate research and not participating in gift economies directly related to corporate goals. However, in today's landscape of life science research, biological innovation takes place in increasingly complex and mixed configurations, in which open data policies and open access coexist with different, and more strict sets of intellectual property rights or secrecy. Further, the corporate world has increasingly been using diverse and mixed approaches to intellectual property, and in some cases – such as database management – strictly proprietary models are seen as no longer sustainable.

However, while Capua never specifically refers to hacking, her plan of action and her public image present references to the digital world and are indebted to open source software and hacking traditions. Capua is hacking biology and its rules because she is actively participating in the transformation and shaping of the current platforms for genetic data sharing, as well as the institutional configuration that sustains it. She is neither merely another rebel scientist or a prominent member of the movement for open access in scholarly publishing. Again, with her attack against a 'closed' public institution Capua contributes to changing the information environment towards a more open ecology.

Her attack is in the name of openness and sharing, but also in the name of the need to subtract power from the slow, non-transparent and corrupted priesthood which runs Big Bio in the public sector. The GISAID case shows one of the many facets of open science politics, one in which openness and transparency are weapons marshalled against the overwhelming power of bureaucracies and can be used to correct an imbalance of power. In this case, what seems to be at stake is scientists' direct and non-mediated power over information management and sharing.

Yet the liberal morality interpreted by Capua is not related to freedom of speech but rather to individual freedom from state control and, of course, to the technical advantages that open science might guarantee to life sciences research. Also, as a result of her 'rebellion', new scientific and media alliances were formed. In fact, GISAID was founded as an independent database sustained by a new hybrid network of actors: scientists, public institutions, private companies and media; all of whom participated in its creation. Thus, institutional change towards open science policies was Capua's most visible result among many.

6
We are the Biohackers

DO IT!
  Jerry Rubin, 1970

Some people actually call themselves *biohackers* and refer explicitly to the hacker movement and history. Their approach towards the life sciences is one that adapts hacker practices to the hardware needed in labs and to the *wetware* that represents the material of life sciences. The kind of open science they foster is one in which openness is not limited to open information sharing, but rather expresses a radical request for opening science's boundaries which allows entry for people who do not belong to its institutions. More interestingly, these biohackers represent all the complexity and heterogeneity of hacker politics, and translate it into the world of biology.

DIYbio (Do-It-Yourself Biology)[1] is a network of amateur biologists established in Boston in 2008 and today composed of several groups and communities in major US, Asian and European cities. Their aim is to provide non-expert, citizen biologists with a collective environment and cheap and open source tools and protocols for biological research which can be conducted in amateur settings. This so-called 'garage' or 'citizen' biology is conducted in weird places such as garages or kitchens and ranges from high-school-level educational experiments to complex biotechnology projects put into place outside institutional settings such as university or corporate laboratories.

During the last few years DIYbio and other related groups have become a distributed movement that is spreading outside the US, and reaching other countries and continents. It has also attracted the attention of the media, who have covered the birth and the evolution of DIYbio and other related citizen biology projects with some intensity.

In one sense, citizen biology is part of a well-known story: the emergence of online platforms for the open and collaborative production and sharing of information and knowledge (Benkler

2006). In fact, DIY biology is based on the same premises that allow for the existence of online distributed social production: cheap and diffused hardware connected to a distributed network (the Internet); collaborative software tools and services; broad availability of data and information that are easily accessible and in the public domain; copyleft licenses that allow content reuse, modification and redistribution; and a culture of participation. Within this general framework, in the last few years we have witnessed the emergence of science movements that rely on the Internet in order to share data and information and to organise offline groups that are geographically dispersed (see Chapter 2). Yet DIYbio and other similar projects represent a shift from bits to atoms, as citizen biology practices are based on building, sharing, cooperating on and creating with material objects rather than mere information.

In a sense, these movements represent today's expression of an old phenomenon, as the free software movement can be seen as the latest expression of a long history of the participation of carpenters, mechanics, miners and outsiders in knowledge production (Conner 2005). It is not difficult to imagine including citizen biology in this narrative. On the other hand, citizen biology represents a new type of challenge to Big Bio as well as a new way of interacting with the system of mainstream biomedicine. These experiences represent an open and collaborative science not limited to the expert community, but rather one that opens and crosses the frontiers of expertise and scientific institutions, fuelled by a hands-on approach to biology.

In fact, the diffusion of the collaborative web and transformations in the way science is conducted have given people new tools that allow for a proactive approach to scientific knowledge and information production, and to the shaping of the technoscientific environment in which they live. This makes DIYbio a very interesting example of a direct translation of free software and hacking practices into the realm of cells, genes and labs.

Thus, in this chapter, my reference to hacking can be more direct, as most of these amateur biologists are eager to be called biohackers. In fact, besides the very diverse cultural genealogies and individual histories that make up citizen biology, the members of DIYbio also have straightforward relationships with the hacker movement. For example, their models are hackerspaces: collectively run spaces that are now widespread in Western and Asian countries and where people gather to hack, talk about and work on computers; spaces where community members who share the same political approach to computers, or where subscribers, for a low individual monthly rate, can find computers, tools and other people interested in hacking. Sometimes, when they cannot open their own labs, DIYbio groups collaborate directly with existing hackerspaces in order to set up small labs, or 'wet corners' within the computer hardware that fills urban hackerspaces. DIYbio members and groups are also immersed in a dense entrepreneurial environment where start-ups and new open science companies try to navigate their way through the dominance of the Big Bio market.

I am aware that what we have witnessed represents nothing but a preliminary phase in the development of a possibly broader and stronger movement. Furthermore, in this chapter I focus mainly on the early years (2008–2010) of DIYbio, which represent only a limited range of examples. DIYbio has been the starting point of a broader phenomenon, as today we are seeing the emergence of several citizen biology groups and projects that are independent from DIYbio. Also, my work is based on the emergence of DIYbio in the US, where it is rooted in some peculiarities of American innovation and political culture, while in Europe new community labs and biohacker spaces have different genealogies. Finally, and significantly, so far no important scientific innovation has come from citizen biology. My point, though, is not an analysis of the scientific potential of citizen biology, but rather of the political and societal novelties that characterise it.

DIY biology has been described in terms of open and peer knowledge production, a danger to public health, a co-optation phenomenon, a democratic (or apocalyptic) change in the relationship between experts and non-experts, an ethical dilemma and an experiment in public engagement with science.[2] Yet what is interesting here is the possibility of drawing a comparison between citizen biology and the history and culture of hacking and free software and the way it represents an updating of the Mertonian ethos. In this sense, the entrepreneurial practices, the relationship with institutions and the approach to intellectual property that characterise this movement are key components of DIYbio's culture.

This allows me to rethink the meaning of DIYbio's ideas of participation in science and to suggest that this movement is an actor in shaping the current innovation regimes in the life sciences as well as playing a role in the relationship between research, academia and the market. In fact, citizen biology represents both an intervention in the marketplace and an alternative mode of scientific knowledge production. Its genealogy is rooted both in countercultural critiques of capitalism and in a corporate ideology of distributed creativity.

It participates in the defining of new forms of science commodification while at the same time it gives people new tools with which to reappropriate and call into question biological research.

This chapter is based on four months of participant observation, both online in DIYbio mailing lists, and offline within local groups and hackerspaces on the US West Coast, particularly in Los Angeles (with SoCal DIYbio) and Seattle. I also conducted interviews with several members and analysed two years' worth of communicative materials, starting from DIYbio's founding in 2008 to the expansion of its network in 2010, which were taken from multiple sources such as press articles, interviews, scientific papers and groups' websites.

Early specific references to the possibility of enacting a biohacker way of conducting life sciences research can be traced back to the Critical Art Ensemble activity in the early 2000s. In its 'free range grain' public performance, for example, this artists' collective based in the US allowed the public to test foods through basic molecular biology techniques in order to find out whether there was contamination from GM crops and thus contest the food trade system.[3]

A few years later, in 2005, Rob Carlson, a physicist who works in the field of genetics, wrote in a *Wired* article that 'the era of garage biology is upon us', linking the participatory culture of web 2.0 to biology, based on part standardisation and extreme informational metaphors (2005). Carlson was working at a Berkeley lab and got inspired by the history of the computer revolution that had happened 30 years before in San Francisco Bay Area garages – the connections between garage biology and hacker myths could not have been more explicit (Golob 2007; Ledford 2010). Three years later, in the other epicentre of hacking history, DIYbio was born.

In fact, the movement started in Boston in 2008 stemming from an idea by Mackenzie Cowell, a young web developer, who was soon joined by Jason Bobe, the director of community outreach for the Personal Genome Project at Harvard Medical School. At the first public meeting, held in a pub in Cambridge, Massachusetts, 25 people turned up. By 2010 about 2,000 people had subscribed to the DIYbio mailing lists. As of 2012, thousands of amateur biologists have been involved in DIYbio activities, even though only a small number of them are active. Yet already at the end of 2010 DIYbio counted several local groups, with new chapters popping up in places as far from Massachusetts as Madrid, London and Bangalore.

Over the last two years the movement – not always specifically linked with DIYbio but inspired by it – has expanded to several

European metropolitan areas and some Asian and Oceania cities.[4] In the US big groups are based in the San Francisco Bay Area, New York, Los Angeles, Chicago and other metropolises. DIYbio is not a formal organisation but rather an open brand that anyone can use for citizen science projects, coupled with a global mailing list where most discussions are conducted and decisions taken. In collaboration with, or partially overlapping, DIYbio, several other citizen biology projects have emerged and form a complex network of different experiences. Still, some more visible members somehow have the ability to direct this brand and are thus identified with DIYbio.

I want to highlight that in many cases citizen biology consists of very elementary scientific practices, and community labs are often poorly equipped and cannot be compared to corporate or academic labs. While over the last few years citizen biology has benefited from the spreading of more open source hardware for biological research and from an increased circulation of knowledge within communities, DIYbio activities often consist of basic practices such as DNA extraction or bacteria isolation with household tools and products (you basically need a kitchen centrifuge, dish soap and a few other easily available chemicals to create a buffer solution and extract DNA from strawberries).

More advanced projects include basic DNA cloning with polymerase chain reaction technologies, gel electrophoresis, or the production of genetically modified bacteria through recombinant DNA – for example, in order to develop glowing, fluorescent *Escherichia coli*. In some cases, the media attention has overstated and mythologised poor scientific practices: right now citizen biology is not a site of research and innovation but rather of political, artistic and educational experimentation. But during the first few years since its founding, DIYbio groups have started several scientific projects. Interesting projects have focused on building open source lab hardware. The Pearl Gel System is a cheap and open source gel box that can be used to run electrophoresis. One DIY biologist has created the Dremelfuge, a centrifuge that works with a cheap and very diffused power-tool gadget. The design for the centrifuge can be downloaded for free and fabricated with a 3D printer (Ward 2010). In the BioWeatherMap project, people were asked to collect bacterial samples from crosswalk buttons in their cities in order to analyse the geographic and temporal distribution patterns of microbial life in a highly distributed way. Other groups were planning to use Amazon's cheap cloud computational power and JCVI Cloud Biolinux software in order to conduct grassroots bioinformatics and data analysis. In New York, DIY biologists were extracting and genotyping people's DNA at public events.

In the past years DIYbio has also established dialogues and relationships with universities, private companies, media and the US government. DIYbio has raised concerns of security and safety among biologists, ethicists and government agencies (Schmidt 2008). This is why the movement has an intense relationship with the FBI and with the Presidential Commission on Bioethics. After the problems in the US caused by people who performed citizen biology post-9/11, and the anthrax hysteria, both the government and DIYbio want to prevent possible problems, misunderstandings or surprises. In fact, one of the images the press uses to talk about biohacking is that of biosecurity and even bioterrorism: are crazy kids playing with dangerous bugs that some terrorist might use to spread unknown diseases and panic (Schmidt 2008)?

Hacking has a dark face that is related to data and identity stealing, cracking, virus production, and so on. Indeed, 'hacking is good. But you have to admit the word has a bad reputation', as a *Nature Biotechnology* article argued (Alper 2009, p. 1077). Furthermore, DIYbio has appeared in dozens of media reports in newspapers and magazines such as *The Guardian, BBC, The New York Times, The Boston Globe, The Economist, Wired* and the like. Also several mainstream scientific journals have covered the DIYbio rise – for example, *Nature* and *EMBO Reports* (Alper 2009; Ledford 2010; Nair 2009; Wolinsky 2009).

Through their website and several local online spaces the members of DIYbio organise collaborative research projects and share scientific data and information. The people who compose DIYbio are very diverse, and they basically belong to three different groups: young biologists, such as graduate or even undergraduate students; computer scientists and geeks who want to tinker with biology; and bioartists interested in applying the critical approach of DIY to biology.

Some members are concerned with the fact that no real garage labs are in place and that access to biological tools and lab equipment is hard to get, expensive and strictly regulated, so that a real DIY biology movement is far from appearing. Yet DIYbio and other citizen biology projects opened several community spaces, such as Sprout in Massachusetts and GenSpace in New York and launched BioCurious, a community lab that was opened in Silicon Valley in 2011.

HACKERS, REBELS AND CITIZENS

DIYbio is often referred to as a biohacker community, and its members use that type of definition very freely. The answer to the question 'Who is a biohacker?' in the DIYbio FAQs includes: hacking as a subculture; the combination of the hacker ethic of 'biologists, programmers, DIY enthusiasts'; explicit references to the Homebrew Computer Club and the Free Software movement; the importance of enjoying 'hacks; and the 'biopunk' attitude.[5]

One of the big public events that presented DIYbio to the world was the hacker conference CodeCon, that in 2009 replaced one third of its normal program with a special focus on biohacking.[6] The media narratives on DIYbio use the definition *biohackers* almost ubiquitously, together with similar definitions such as *life hackers* (Ledford 2010), and they often draw comparisons between citizen biology and the Homebrew Computer Club, the headquarters of 1970s Bay Area computer hackers such as John Draper (Captain Crunch), Lee Felsenstein and Apple founders Steve Wozniak and Steve Jobs (Bloom 2009; *Economist* 2009; Golob 2007; Johnson 2008).

Yet some individuals linked to DIYbio prefer to define themselves as makers, craftsmen, enthusiasts, hobbyists or amateurs. They often agree, though, that the garage is an important part of the love the media express for DIYbio. Community labs are places where one can develop his/her curiosity, creativity and desire to tinker with genes and cells. After all, hackers that performed the computer revolution were nothing but 'a bunch of unshaved guys in a garage' (Golob 2007).

Press accounts of DIYbio and the members themselves highlight how garage biology is to be considered part of the tradition of American innovation – think about Apple or Google and the mythology related to the Silicon Valley garages where they started operating (Levy 2010a; Vise and Malseed 2006). In public discourse about biohacking it is not uncommon to run up against statements such as 'the future Bill Gates of biotech could be developing a cure for cancer in the garage' (Wohlsen 2008).

Other similarities between DIY biology and hacking are in the obstacles biohackers identify in Big Bio. As I will show below, in DIYbio narratives universities and corporations are flawed because they rely on high levels of specialisation and hierarchical systems, but also because they build monopolies and steal individual creativity by means of intellectual property rights. The Hulking

Giants of Big Bio are neither open nor inclusive. Perhaps, as DIYbio founder Jason Bobe said, 'there will always be the giant players – the biotech and pharmaceutical companies – in life sciences' (Nair 2009, p. 230), but the widespread diffusion of information and sequencing technologies will allow amateur biologists to contribute to the scientific enterprise.

Of course, fun and hedonism are also important ingredients of the DIYbio culture that amateur biologists find difficult to express in corporate and academic labs. As the DIYbio cofounder Mac Cowell said, DIYbio 'gives people the justification for doing silly or weird things' because, as in many narrations on rebel science and on hacking, 'innovation arises from having fun and playing with biology'.[7] Cowell quit his job because 'he wasn't having fun anymore' and sold his car to start DIYbio (Boustead 2008). Just as Wozniak sold his HP calculator and Jobs his Volkswagen bus to start Apple in Jobs' garage (Levy 2010a).[8]

Hackers do not always like the sunlight. On the ninth floor of building 26 at MIT, hackers would work all night in order to avoid the priesthood which wasted precious time using university computers for what hackers perceived as dumb tasks, but also because of their weird circadian rhythms and lifestyle. And so do biohackers: 'you'll be tweaking genome sequences on your computer late at night' (Carlson 2005). You won't be able to stop the passion of hacking. DIYbio discourse is also laden with informational metaphors: the standardisation, abstraction and digitalisation of genetics will give more people the opportunity to do biology (on this see Haraway 1991; Kay 2000; Keller 2000).

The DIY side of citizen biology is also related to a turn in life sciences towards more active and interventionist practices, recently enacted by the rise of synthetic biology and its accent on *making* organisms rather than modifying them (Calvert 2010; Keller 2009). Yet the use of software metaphors is not limited to genes. When dealing with the FBI, biohackers want to highlight how and why they want more transparency: 'it is like software ... it is security through transparency'.[9]

References to hacking are dominant. But the use of the term 'do-it-yourself' positions DIYbio within an old American movement of makers and inventors who work in their garages, and also gives it a rebel flavour. The expression 'DIY' was broadly adopted in the 1980s by the punk-hardcore movement both in America and Europe. Yet this movement is witnessing a renewal and is now part of a broader social phenomenon centred around the convergence between online peer production, the diffusion of cheap and open source tools and machinery (such as 3D printers) and a widespread 'maker' culture.

DIYbio is part of this movement, whose main communication tools are magazines such as *Make* or websites such as *Instructables* (Tocchetti 2012). This is also a link between biohacking and craftsmanship. Christopher Kelty, one of the first scholars to address DIY biology from the point of view of its societal and cultural dimensions, argues that three figures can be used to understand citizen biology, namely outlaws, hackers and Victorian scientists (Kelty 2010). DIYbio, in some media accounts, is 'a throwback to the times when key discoveries were made by solitary scientists toiling away in their basement labs' (Nair 2009, p. 230). One of the founders of DIYbio, Jason Bobe, also draws this comparison:

> in some sense, we're returning to some of the roots of biology, where scientists had laboratories in their parlors. You know, it was parlor science. It was something that didn't actually happen often in institutional settings; it was something that happened at home. (NPR 2009)

For Drew Endy, a Stanford bioengineering professor who is one of the strongest backers of the DIY biology movement, 'Darwin may have been the original do-it-yourself biologist, as he didn't originally work for any institution' (cited in Guthrie 2009).

Thus, for its members, DIYbio is not only biohacking but many other things as well. It is public engagement with science, open source, decentralisation, participation and innovation. When asked to interact with the FBI or with the US Presidential Commission for the Study of Bioethical Issues, DIYbio proved very capable of finding ways to position itself in order to avoid problems and any backlash. For example, they decided to highlight that citizen biology has an educational side and that it could provide cheap hardware or kits to be used in schools or community labs, besides giving people a vibrant online community to discuss science. DIYbio might become a 'cultural interface' for biology, a place for people to explore biotech. In their letter to the Presidential Commission, members argued that:

> DIYbio.org was created to help build a positive public culture around new biotechnologies and practices as the number of

contributors to the life sciences extends beyond traditional academic and corporate institutions. (DIYbio 2010)

There is a classical problem regarding the relationship between science and society at stake: distributed participation. It is easy to state that online commons-based peer production practices are changing, and increasing the ways of participating in the production of scientific knowledge. DIYbio contributes in adding physical spaces and material production to online peer production. But does this increase consist of a real shift towards democratising science? Does it actually affect the asymmetrical relationships between citizens and experts?

Scholars who have tackled this relationship have generally been very prudent in picturing participation in science. Often, ambivalence is highlighted. Callon and Rabeharisoa (2003) point out that 'research in the wild', or the intervention of patients in biomedical research, involves their active participation in establishing new collectives that include new subjects. Also, the renegotiation of the relationship between research in the wild and research conducted in professional settings involves issues of power, epistemology and the presence of incentives of a new and different nature.

The changing panorama of expertise urges lay people to get actively involved in technoscientific decisions in order to change the world and not just observe it (Collins and Evans 2007). While referring to geeks and the diffusion of free software practices outside the computer world, Kelty (2010, p. 8) argues that the public can be 'aggressively active' instead of simply not being passive. Do-it-yourself science certainly challenges mainstream science, asking for more access and involvement. But amateurs are also redefining what being 'the public' means in the current configuration of science–society interaction: an active role substitutes the simple encounter between science and its public and creates new spaces of interaction and participation (see Nowotny 1993). DIYbio is a site where different approaches coexist. For example, through DIYbio amateurs who work outside of traditional professional settings can have 'access to a community of experts'.[10]

This is not too different from the perspective of Critical Art Ensemble (CAE), an art/activist group whose works and writings are considered by many DIY biologists as a foundational story. In 2004, one of the CAE members, Steve Kurtz, was arrested under the suspicion of bioterrorism when, after his wife died of a heart attack, the FBI found cell cultures and lab equipment in his apartment (Simmons 2007). CAE used amateur biology as a tactical practice in an artistic context in order to create what they called 'a countersymbolic order' against the power of Big Bio. The public space their practices aimed at creating was intended to be

> one where the authority of the scientific personality is not so powerful. The hierarchy of expert over amateur has to be suspended in this context. If experts have no respect for the position of amateurs, why would they come to a place where dialogue is possible? (Critical Art Ensemble 2002, p. 66)

Yet the vision of citizen biology as a site for participation has a completely different side. While for CAE the goal was to enable people to challenge the capitalist face of Big Bio by providing conceptual and political tools, in some biohackers' views participation could help overcome some of the problems faced by Big Bio itself. In fact, biologists are gathering more and more genetic data without knowing exactly what to do with it. For citizen biologists 'genomes are useless right now. They can be useful if people share their phenotype, and that is something the citizen himself has, not the expert. The future of human genomics will depend on individuals sharing'.[11]

These claims of course resonate very well with direct-to-genetic companies which exploit gift economies by providing sequencing services coupled with online sharing platforms, such as 23andMe (Levina 2010, see Chapter 2). The citizenship imagined by DIYbio keeps together the necessity of sharing information in networks where private appropriation occurs, as well as the will to enact forms of open data sharing alternatives to the intellectual property rights enclosures that sustain Big Bio monopoly power.

FREE AS IN FREE GENES

When it comes to openness and sharing, DIYbio members would certainly agree with the free software foundational definition: 'free as in free speech, not as in free beer'. Access to knowledge is another important framework under which DIYbio operates, as it enables citizen participation in science. Indeed, openness is one of the core legal implications and needs of user-led science. Citizen science and public involvement, and collaborative models between scientists and non-scientists, need policy solutions that support not only data and knowledge sharing, but also the sharing of the different types of benefits deriving from it. The incentive model of citizen science is closer to that of open source software than to that of Big Bio (Stodden 2010a). But for DIYbio, openness refers both to the open access to data and knowledge according to an explicit open source model, and to open participation directed to *all*, regardless of professional recognition from Big Bio. As DIYbio wrote in its online FAQ page, the organisation is a 'groundwork for making this field open to anyone with the drive to become great at it'.[12] In which sense, then, would the free software model apply to genes and cells?

DIY biologists have different modes for finding the tools and machinery needed for their labs. These tools are usually very expensive or difficult to buy since companies do not often sell equipment, reagents and so on to individuals – mainly for safety and regulatory reasons, but also because, as one amateur biologist says, 'they do not perceive the possibility of a non-institutional market', which constitutes a threshold that is hard to overcome. The story of two polymerase chain reaction (PCR) machines can explain how DIYbio answers this problem.

In San Francisco, two young electrical engineers, Tito Jankowski and Josh Perfetto, have developed OpenPCR, a project to build a cheap PCR machine under open source principles: anybody is free to download the instructions to build it and the software to run it, and then have an easy to use, $400 machine at their disposal. As for other DIYbio projects, the money needed to develop OpenPCR has been raised with a crowdfunding scheme through the website Kickstarter. Yet in Los Angeles, SoCal DIYbio have found two used – and broken – PCR machines that the group fixed using members' electrotechnical skills and adapting free software to make it control them.

Other DIYbio techniques for putting together cheap equipment include stealing, buying used stuff such as benches or glassware from university labs, or using the university address of their graduate student members in order to get material shipped from companies. They also use skills some of them acquired working in 'ghetto labs' in universities that were not well funded. As a DIY biologist said, 'we don't care where the shit comes from. We want shit that works!' Before receiving a PhD at an important US university, a DIYbio member had worked as an undergraduate in a small lab, where she:

learned some skills on how to run a lab without spending any money: how to get free equipment from companies asking for sample ... I have a ghetto sense now. But still, we published papers out of that lab.[13]

Again, apart from the problem of intellectual property, citizen biologists have a complex relationship with big institutions. On the one hand, they rely heavily on universities for materials, education, used machinery and other needs. Yet they also have problems with being recognised as having real scientific projects. In 2009, for example, DIYbio was excluded from the annual iGem competition, a global synthetic biology contest based at MIT in Cambridge, Massachusetts, in which teams of undergraduate students from all over the world compete to design and build the best biological systems and operate them in living cells. The starting material is composed by the items collected in the Registry of Standard Biological Parts. In 2012, for the first year, iGem was open to entrepreneurs and started considering proposals related to new business and market models for synthetic biology.[14]

An interesting feature of DIYbio values is that often intellectual property rights are not perceived as evil per se. It surely adopts a very open attitude, using open access tools and copyleft licenses when it comes to sharing data, protocols and knowledge. Yet the ambivalence that characterises DIYbio emerges when, talking to different members, one finds out that for some there is a political commitment to open science as a form of free speech comparable to the liberal vision embodied by free software: to prevent people from making science is against freedom of thought, and access and transparency would in fact enable people to develop a more proactive attitude towards their own biomedical information.

Hackers have always adopted strategies against closed protocols, and over the last few years biomedical information has become one of the targets of such practices. In September 2012 the Italian hacker and artist Salvatore Iaconesi announced through an online video that he had brain cancer. But the medical information and data he was given at the hospital, such as brain images from magnetic resonance and computed axial tomography, was in a format he couldn't access. So he publicly announced: 'I cracked them. I opened them and converted the contents into open formats, so that I could share them with everyone', and then put all the data online for physicians, artists and activists who might give him an 'open source cure' in the form of an artwork or a map based on his data, or a solution for his health problem. Iaconesi's hack sparked a broad debate over access to medical data, pushing

Italian institutions to consider making the use of non-proprietary formats for digital medical information compulsory.[15] But while this example of hacking directly represents the will to enact free-speech-like openness, in other cases openness is a means towards a different end, such as new forms of entrepreneurship.

In a similar way to the passage from free software to open source (see Chapter 2), openness can be a way of defying incumbents and restoring the freedom of the market. 'If only I could have put all the money I paid the damn lawyers into the molecules!' stated a biomedical entrepreneur who had to overcome the obstacles represented by the broad patents owned by Big Bio[16] – a typical anticommons effect (Heller and Eisenberg 1998, see Chapter 1). Thus, often when DIY biologists talk about innovation happening outside traditional settings such as academies and corporations, they also want to highlight that openness is not only good per se but rather part of the strategies against Big Bio monopoly power. Indeed, biohacking is laden with anti-institution and anti-bureaucracy claims.

Giving people cheap and widespread tools for biology, some DIYbioers want to 'make people less reliant on other people for living a good life'.[17] They also want to avoid academic paternalism and demystify 'official' science. For example, even though many members are getting their PhDs, the importance of the normal, institutional course of scientific education is not taken for granted. One important barrier to entry for people who want to practice biology is formal education: a PhD title 'is over glorified and I want to show it' said a graduate student convinced that participation in DIYbio projects was more important than a formal, 'normal' university career,[18] something biohackers want to demystify (Wolinsky 2009).

According to Jason Bobe, we are going to see a scientific renaissance that will be funded and enacted outside the incumbents of Big Bio and their slow and bureaucratic processes. The peculiar feature of this renaissance is that 'it's going to take place outside of "science proper", away from universities which dominate now, and funded out-of-pocket by enthusiasts without PhDs' (Bobe 2008). In sum, 'we're all doctors here, man!' as another DIY biologist said during a meeting.[19] Moreover, formal education is an aspect of Big Bio that some citizen biologists cannot stand because it is the expression of the power of an old-boy network that must be broken: 'nowadays, biology is like a medieval guild. Firstly, you have to get a PhD, but if you want to practice then you need venture capital, otherwise you don't have the tools' (Bloom 2009).

Autonomy is at the centre of the vision expressed by the founders of the Silicon Valley biohacker space BioCurious, according to whom people 'want a space where they can work on their own projects, outside institutions they hate, such as universities and corporations'.[20] Other DIY biologists are even more enraged: 'the Bayh–Dole Act sucks! People don't want to give their ideas and their intellectual property to their institutions'.[21] Patents are not only a moral problem or an obstacle to innovation, but something scientists should be protected from (and not with). The Bayh–Dole Act urges researchers working in public institutions to patent their findings and inventions through the university (see Chapter 2). But for some DIY biologists this is a problem related to individual rights rather than a more general problem of knowledge privatisation and academic capitalism. They want to keep their intellectual property rights and not to remise them to the big institutions they work for in their daily jobs.

THE HOMEBREW MOLECULAR BIOLOGY CLUB

DIYbio embodies all the different faces of hacking such as openness in data and knowledge sharing as well as the openness of the doors of scientific institutions, but also rebellion, hedonism, passion, communitarian spirit, individualism, entrepreneurial drive and distrust for bureaucracies. DIYbio is a really interesting case because it includes all the cultural and political complexity of hacker ethic and FLOSS practices. In this case, rather than highlighting a peculiar recombination, it is more correct to talk about the coexistence of several cultural traits taken from different ethical sets, such as the Mertonian ethos, the hacker and free software cultures, and older cultures such as that of the Victorian gentleman scientist.

Biohacker spaces are meant to be directed towards partially overlapping but also radically different goals. Silicon Valley's BioCurious is meant not only as an educational space but also an incubator for Bay Area entrepreneurs that need a coworking space, and often hosts courses on biotech business models. The Los Angeles biohackerspace seems to emphasise freedom and autonomy from academic institutions. La paillasse, the Paris community lab, is directed towards civic goals, as it claims that 'citizens must have in their hands a counter-power to participate in the societal choices concerning the use of these technologies' (Meyer 2012).

For now, the results of DIYbio have been modest. From a scientific viewpoint it is hard to state that they are actually hacking DNA and cells, and we do not know if they will be able to hack them in the near future. Yet DIYbio is making biology hackable in other ways. First, the kind of acknowledgment and incentives they recognise are not always related to the ones of institutional science: a good hack does not need to be peer reviewed, although it surely has to be shared with other biohackers. You do not need a PhD to do biology.

Second, citizen biologists use informational metaphors and aim at standardising genetics in order to make it cheaper and more easily accessible through open source approaches. Third, they are opening community spaces for people to conduct biology outside the boundaries and limits of Big Bio. Finally, they are trying to open up the boundaries of life science entrepreneurship by experimenting with new business models based on open source approaches.

Yet regardless of its poor scientific output so far, DIYbio's success seems to be rooted in its symbolic power. DIYbio, making biology hackable in all these different ways, is producing the picture of a different way of conducting research in the life sciences: more open, horizontal, within a very mixed constellation of different actors such as start-ups, universities, individuals and community spaces, with a prominence of small and open companies instead of the slow giants of Big Bio.

They are also a powerful antidote against the critiques that have hit biology during the last 20 years, after the wave of privatisation and patenting, and the transformation of institutional settings in which life sciences are conducted. Thus Big Bio will perhaps have to take their needs and interests into account. Companies and scientific institutions are asking citizens to contribute by crowdsourcing knowledge, sharing and analysing data, or performing scientific research. Will they be able to open themselves up to a more inclusive relationship with citizen science? Well, if they won't, they might have to face rebellion, at least according to some biohackers. In her *A biopunk manifesto* the hacker and DIY biologist Meredith Patterson pompously (and ironically) states:

> We the biopunks are dedicated to putting the tools of scientific investigation into the hands of anyone who wants them. We are building an infrastructure of methodology, of communication, of automation, and of publicly available knowledge. ... We reject the popular perception that science is only done in million-dollar university, government, or corporate labs; we assert that the right

of freedom of inquiry, to do research and pursue understanding under one's own direction, is as fundamental a right as that of free speech or freedom of religion. ... The biopunks are actively engaged in making the world a place that everyone can understand. Come, let us research together. (2010)

With its radical requests for openness and inclusion and with its rejection of institutional prerogatives and constraints, DIY biology surely challenges many assumptions about public participation in scientific knowledge production. Citizen scientists and users contributing to science claim to be part of the scientific process on almost any level.

Like other hacker communities and practices, they point out a problem in the current dynamics of power over information and knowledge production. Distributed social production has already proven to be enormously productive in many fields of human knowledge and DIYbio claims to make a positive change accompanied by a redistribution of that power. Yet for some DIY biologists, the relationship with institutions and Big Bio cannot be one of a punk refusal, a menace against the established order. DIYbio needs to be transparent, friendly and open to dialogue: 'we want to encourage people not to be punk, underground biohackers.'[22] Thus citizen scientists depend on big science but try to live beyond its frontiers, in no man's land: they are somehow *outlaws* (Kelty 2010). This has important implications for the relationship between different types of expertise.

In more than one aspect, DIY biology's evolution is similar to other forms of knowledge production situated outside the boundaries of institutions. In *Convergence culture*, Henry Jenkins (2008) depicts the clashes that involve fans and mainstream media industries. Fan creation 'in the wild' – to connect Jenkins with studies on the participation of lay people in biomedical research – can be a very rich field that companies can harness to capitalise on new content and get in touch with their public (Terranova 2000).

On the other hand, companies are always challenged by content creation that happens outside their boundaries because they need to control it in order to avoid injury, and this can be a very expensive and puzzling task. The pessimistic side of this balance is represented by the exploitation of people's creativity and the appropriation of free labour by greedy corporations (see Chapter 2). This description somehow echoes Marx's ideas of the relationship between capital and labour. Italian autonomist Marxists have, since the late 1950s, argued that workers' struggles are among the main engines of technological innovation and of capitalist transformation and evolution. Yet capital is never able to fully control workers' social practices, or to reconcile its inside and its outside. This edge is where capital struggles to survive, feeds on new ideas and solutions, and therefore evolves. Struggles against exploitation are both the driving force and the opposition of capital (Hardt and Negri 2000; Marx 1990; Panzieri 1976).

Of course, DIY biologists are neither workers struggling against capital, nor fans shooting a short movie of the Star Wars saga without the authorisation of George Lucas, but they have an ambivalent role with respect to Big Bio. One interesting question is whether in the future their hacks will favour, change or disrupt today's life science incumbents. They challenge the separation between the roles of experts and non-experts in new and deeper ways. They refuse the absolute authority of universities on scientific recognition (we're all doctors here!) and of both academia and industry on intellectual property rights (the damn lawyers! The Bayh–Dole Act sucks!). But they also represent an attempt at new ways of participating in an innovation regime that includes universities, corporations, start-ups, patients' associations and so on. Many members of DIYbio refer to, for example, the possibility of developing a new market for biology tinkering tools or for small companies not dependent on patents but based instead on open science practices – 'we gotta find a way of marketing this!'

Less than ten years ago, CAE was highly sceptical about the possibility of a corporate side to amateur biology when it argued that 'even entrepreneurs do not seem to have any interest in finding a way to capitalize on this divide' between experts and amateurs (Critical Art Ensemble 2002, p. 123). Yet an important part of DIYbio invests precisely in the role of entrepreneurship and corporations in sustaining a possible biohackers movement. In this sense, again, DIYbio's relationship with Big Bio is complex and ambivalent.

The anti-bureaucracy side of DIY biology is determined to challenge Big Bio incumbents. Yet many self-described biohackers collaborate with or are employed by academic and corporate laboratories. Their autonomy from institutions is questionable, as several are part of the legitimate biomedical research system by day and in their spare time pursue their personal agendas within the freedom created by DIYbio practices. In their discourses, amateur biologists would like to dismantle monopolies based on intellectual property rights, capital-intensive laboratories and scientific expertise. Yet most of them are not interested at all in a critique of academic capitalism or biocapitalism. Many are looking for fun and like to tinker with cells and genes, and might view citizen biology as an educational practice. Some are interested in the possibility of opening up new markets where smart, small scale and open source models could compete with Big Bio and its Hulking Giants. Others hope Big Bio would finance their activities, recognising biohackerspaces and biohacker communities as innovation incubators where new ideas, start-ups and entrepreneurs might come from in the near future.

Both models are similar to the free and open source software economic models. On the one hand, a challenge to big, closed corporations enacted by small and peer production projects that can count on openness to harness people's cooperation and find faster and better solutions. On the other hand, a direct participation of big corporations in open production.

While we cannot predict the scientific and political outcomes of DIYbio and citizen biology, the interesting feature that characterises this movement is that it embodies many parts of the hacker political spectrum. Very different versions of liberalism are at work and coexist in DIY biology, as well as the re-emergence of a pre-modern scientific ethos. DIYbio updates and transforms the cultures upon which biocapitalism is based, by incorporating elements such as openness, flexibility and distributed participation. Furthermore, these biohackers give new meaning to hacking itself, as they expand its frontiers into new realms outside the traditional boundaries of computer science.

7
Conclusion: How to Hack Biology

There's a place for you in the New Information Order.

Processed World, 1982

Hacking biology is different from hacking DNA. Hacking biology means to change and recombine technoscientific cultures and science institutions rather than genes and cells. Over the course of this book we have seen examples of an innovative recombination of cultural elements that characterise an emergent part of the life sciences system. The multidimensional and diverse elements of hacker cultures represent the cultural material that some biologists are using in order to renew the toolkit of strategies of action they have at their disposal.

The result of this process is a new open science culture related to transformations in the way biomedical research is conducted, shared and appropriated. The biologists I have presented here are not merely open access and open source advocates. My proposal is to consider this remix between the Mertonian ethos of twentieth-century scientists and the ethic of hackers as a new phenomenon that not only embodies elements related to openness and sharing, but is rather a more complex recombination in which other characteristics emerge alongside them: anti-bureaucracy rebellion, extreme informational metaphors, institutional critique, autonomy, independence, a radical refusal of external interferences and also of scientific institutions themselves, hedonism and finally an intense relationship with the media.

This culture expresses the re-emergence of an ancient and recurrent element in the history of science, namely the fight between openness and closure. But the complex and diverse cultural repertoire of the biologists included in this study is different from the classical ethos of modern scientists who work in academia: those who are disinterested, respectful of bureaucracy and peers and not compromised with the market. At the same time, it is also different from a corporate ethos of secrecy, hierarchy and closure. Biohackers are a much more complicated expression of the transformations of the norms regulating today's production of scientific knowledge, a change that affects the relationship between biosciences, society, public communication and the market. They also embody a change in the media landscape. The media have a crucial role in making some scientists 'visible' and giving them an important public dimension. But the rise of new media also represents the technological foundations of transformations in the way knowledge is produced, shared, discussed and legitimated.

Hacking biology includes practices that go beyond laboratory doors. The cases I present here share an open approach to information, but this is not enough. These cases illustrate a tension between scientists' cultures and the transformations of post-genomic biological research. They surely show the different possibilities enabled by open science practices, but also features such as rebellion, anti-bureaucracy and participation that have a crucial role in making DNA something people can hack.

All these biohackers criticise the scholarly publication and peer review system. They all struggle against Big Bio bureaucracies and incumbents. Yet there are differences and peculiarities that are useful in that they highlight the contingent and very diverse typologies of hacking they represent. In fact, they substantially differ from each other – I have included famous and wealthy biotechnologists who have access to capital, research facilities and political power, and a network of citizens who are trying to apply a DIY and low-capital approach to biology.

Craig Venter's *Sorcerer II* made DNA hackable by insisting on informational metaphors, deterritorialising genomes and circulating them in a heterogeneous network of firms, universities, foundations and mass media. Venter bypasses the Mertonian ethos and recombines Victorian gentlemen scientists' culture with a hacker component. He proudly announces that his greatest success is that he managed to get hated by 'both worlds': academic and corporate. Yet Venter's hack is directed towards profit and entrepreneurship, as he tries to exploit openness in order to participate in a different form of biocapitalism in which data circulation is as important as data gathering and management. His shift to open science shows how openness can be part of a corporate (and media) strategy and is directed towards independence, individualism and rupture with the status quo.

With GISAID, Ilaria Capua made the DNA of viruses hackable by removing it from Big Bio's secret world, a world where an old-style priesthood decides who can access the Hulking Giant databases.

Opening up the access to avian flu data was achieved by restoring the modern science ethos and upgrading the image of a rebel and revolutionary scientist with contemporary features taken from the open source and free software world. Her rhetoric is one in which scientists can break the mould and refuse the secrecy of an 'outside world bureaucracy' in order to push giant-sized institutions to change. Here, power over access to data and knowledge (and not money) is at stake, along with scientists' autonomy and civic goals.

The cultural references and practices of DIYbio talk about opening up biology to public participation but also to new forms of grassroots entrepreneurship. Their hacks are not merely a political critique against Big Bio but rather an attempt at finding new and better ways of accessing cells and DNA. DIYbio is a very complex case, and the explicit hacker component it shows encompasses many different facets of hacking: the rebel one, the entrepreneurial one, the anti-monopolist one, the individualist as well as the collective ones. Proactive approaches to technology and critiques of biocapitalism walk hand in hand with the individual necessity of breaking up a monopoly.

In times of crisis and change the need for a reconfiguration of different aspects belonging to pre-existing cultures becomes more insistent in order to answer the urgent need for new strategies of action. Thus individuals can mobilise cultural characteristics and operate a remix between an old culture, already accepted and embodied in a recognised set of practices and norms, and ready to be used; and a different set of cultural features that belongs to other social groups.

The stories I have analysed are powerful precisely because they narrate several possible futures of change, openness and horizontality in a field as difficult and criticised as today's biomedical sciences are. In fact, they show that this culture can be reconfigured in different ways in order to adapt it to different needs. They represent very different worlds, such as academic and publicly funded science, freelance research able to raise money from corporations, governments and venture capitalists, and amateur research that has ambivalent relationships with universities and firms.

Yet putting them together under the umbrella of hacking, I point out the emergence of a new open science culture: a new form of public image that scientists can use to build new strategies of action and better interact with the peculiar socio-economic configuration of contemporary biological sciences. The old Mertonian ethos of the twentieth-century academic scientist is still at scientists' disposal, but in order to use it as a powerful tool it needs to be remixed with components coming from critical cultures directly related to computers and information technologies.

The spreading of legal and technological tools that enact new forms of data and knowledge sharing needs a cultural adaptation that Merton cannot provide. Open science asks for new social, communicative and political practices and a new incentive system. Old media such as peer reviewed scientific journals are not always an adequate answer to new societal and economic needs. In hi-tech gift economies enacted by digital media, data sharing and participation are part of the corporate economic models as well as ways to enrich the commons and challenge monopoly power and its informational land revenues.

Of course, the results from the three cases analysed here are not easily generalisable. First of all, although biology is a very globalised enterprise, research on open science cultures should take into account regional differences: Europe is different from America, which in turn is different from Asia. Silicon Valley is a very peculiar area, where technological cultures develop in their own specific ways. DIYbio is somehow becoming a global movement, but its development outside the US has brought out geographical particularities and increased its diversification.

Furthermore, several other cultures converge in today's open science. Biohacking does not overlap 100 per cent with open biology, which is a much broader phenomenon and shows a more diverse range of practices. Many aspects of the open science movement bear no relation to the kinds of changes I have presented in this book, but are rather the fruits of institutional choices, the need for public control over information, or traditional corporate practices. Biohackers do not cover the whole spectrum of open science politics, yet I think they represent a crucial tendency within contemporary life sciences. They also show that a critical analysis of open science, beyond ideological adhesion or political support as well as beyond the mere opposition to open/closed or proprietary/non-proprietary, is much needed.

While their individual results might not be crucial for the evolution of scientific knowledge, an interesting feature is that very similar cultural configurations can be found in several other biologists working in different institutional settings and also in different scientific subfields such as personal genomics, synthetic biology and metagenomics. During the years in which the events I analysed took place, the scientific and media stages have seen the appearance of other biotechnologists using the same strategies and discourses, exploiting the same images of science, keeping a balance of information disclosure and privatisation, and exploiting the media and the Internet in ways similar to the ones I have described in this book.

Examples that appeared in the post-genomic era (the first decade of the twenty-first century) include several biohackers who can be defined as such according to the meaning I have given to the term. The 'open source junkie' George Church from Harvard, also nicknamed the 'information exhibitionist' given his attitude for total data disclosure, is the director of the open source Personal Genome Project, a long-term project aimed at sequencing and publicising the complete genome and medical data of 100,000 volunteers. Church is involved in many start-ups in the field of personal genomics and participated in the production of the Polonator, a genome sequencing machine that works with open source software, reagents and protocols.

Another is Drew Endy of the MIT BioBricks Project, with his ideas for 'DNA hacking' that he has also presented in public meetings such as the Chaos Communication Congress of Berlin, one of the most famous hacker gatherings on the planet. Endy is among the founders of the Registry of Standard Biological Parts, a collection of standardised genetic parts that can be used in synthetic biology projects.

Church and Endy are two of the most famous US supporters of open genomics and citizen biology. The Icelandic deCODE genetics of Kari Stefansson sells direct-to-consumer genomic services with the motto 'Know your code', in order to discover the secrets of your DNA and 'take a voyage of discovery' through them. 23andMe, the Google genomic start-up, urges you explicitly to: 'Unlock the secrets of your DNA. Today'. But besides cracking the code of your genome, 23andMe asks you to share your genetic, phenotypical and medical data in its social media website. The Spencer Wells' Genographic Project, a massive collection of genomic data started in 2005, is a joint venture between National Geographic and IBM and brings together dozens of universities and research centres from all over the world. This project, half scientific journey, half media production, is based on the selling of a personal DNA testing kit whose results are made publicly available through an open source database.

DIYbio practices have sprung up in both Europe and Asia, with the opening of several biohackerspaces and the emergence of a broader movement of citizen biology projects. These are often related to the art scene but their practices are based on the availability of open source tools for biological research and the sharing of standardised methods and protocols. Among the many possible examples, we could cite European community labs such as BiologiGaragen in Copenhagen or La Paillasse in Paris, or the explosion of life hacking practices in Asia. A project such as the scientist–artist collective Hackteria represents an attempt at creating a transnational network of bioartists based on hacker and open source practices. In all these cases, the informational and promissory metaphors surrounding genomes feed on the changes of a biology in which scientists are managers of open genetic information, providers of customised services and direct interlocutors for the needs of citizens outside of the biomedical industry. They are interpreting the same shifts in the relationship between genetics and society that I have pointed out above. The overlapping of openness, anti-bureaucracy, hedonism and sometimes even explicit references to hacking, is becoming common in today's biology.

Thus I think this emergent class of biohackers is related to a new type of interaction between scientists' practices and biology's social contract, which is at work in corporate and public settings as well as in citizen and patient groups' experiences. A new open science social contract would restore some sharing practices that characterised twentieth-century academic research. But they would be transformed, broadened and improved by web technologies and the widespread diffusion of open and peer production. At the same time, it would include practices of closure such as patents and copyright. Different forms of intellectual property rights would coexist in an environment inhabited by creatures as diverse as companies, universities, public agencies, start-ups and new institutions such as citizen science projects.

Obviously, we can not predict whether this tendency will remain secondary and localised, or if it represents the beginning of a more important and broad change. More importantly, the new open science culture related to this social contract would maintain its political ambivalence, and the direction its evolution will take might depend on political choices more than on cultural change. For example, it is not clear whether biohacking will keep on being dependent on Big Bio institutions for skills, tools, data and even intellectual property rights regimes, or will build an independent movement based on alternative tools and educational methods. Also, it is far from clear what the role of sharing itself will be in future science.

In the twentieth century, science incorporated patents in its cycles of credits thanks to the very active role of scientific communities. Patenting was thus institutionalised as one of academic science's recognised practices. Will the same happen with sharing? Proposals made by the open science movement rely on the idea of accrediting data sharing as a form of publication, thus giving researchers an incentive to share that is related to their academic career and contributing to the institutionalisation of open science.

THE POLITICS OF OPEN SCIENCE

The mobilisations and reconfigurations of the science cultures that I pointed out in this work are going to be more and more important and attuned to the needs of twenty-first-century life sciences. As an initial and simpler explanation, the emergence of similar types of scientists can then be interpreted as a response to a legitimacy crisis.

With their insistence on openness, rebellion and antiestablishment critique, these biologists are making the public image of genetics 'moral' again, an image that badly needs a renewal after the legal, political and societal clashes it suffered because of the rise of the anticommons. In this sense they embody a critical culture opposed to the dynamics of academic capitalism and science commodification. Biology keeps on being a highly moralised enterprise, and this is even more important when it comes to issues of access to information and scientific knowledge production.

This need is not limited to academic scientists: Janet Hope directly links public relations strategies and the use of open source models, the latter being 'an investment in the firm's overall brand and reputation ... a deliberate strategy to enhance the reputation of companies in the biotechnology and related industry' (Hope 2008, p. 262). Also, for Kaushik Sunder Rajan, open sharing of biological data and knowledge can be a strategy of new corporate activism built upon narratives of technologies as forces for positive social change (2006).

But biohackers' mobilisation of different practices and cultural traits can also be interpreted as a way to better interact and position themselves within the current socio-economic configuration of biological sciences. Indeed, biological innovation now takes place in increasingly complex and mixed configurations, in which open data policies and open access coexist with different, and more strict, sets of intellectual property rights, and in which the boundaries of biomedical research are becoming more porous and inclusive. Both academic and industrial research (provided that it is still possible to clearly separate them) have increasingly been using diverse and mixed approaches to intellectual property, and in some cases – such as database management – strictly proprietary models are being outdated by open approaches.

Thanks to the open and free input of voluntary contributors, participatory processes of governance – and the universal availability of the output – open and peer production might prove to be more productive than centralised alternatives. The examples I have provided are not alike in this respect, but represent the growing importance of peer production within open science politics: the stress on distributed creativity that was secondary in the *Sorcerer II* case; increased and became more central in the birth of GISAID, even though it was limited to institutionalised scientific communities; and became one of the main arguments that sustain DIYbio practices, where the building of non-expert citizens' open and distributed participation constitutes the core claim around which communities are organised.

Finally, to continue using the hacker metaphor, these biologists are hacking the rules of biology. Their active approach to information allows them to participate in the transformation and shaping of the current proprietary structure of science. Their struggles against the Big Bio priesthoods are a challenge against the current distribution of power among science's institutions. In this sense, their stories could be a model for changes that are also taking place within other innovation regimes such as software, hardware, technology and so on. In many fields of information and knowledge production, actors are actively transforming and building their own infrastructures – whether they are technological or legal (licenses).

Pierre Bourdieu, while referring to epistemic (and not institutional) revolutions, highlighted that revolutionary scientists not only head towards a victory – they can be willing to change the rules of the game: 'revolutionaries, rather than simply playing within the limits of the game as it is, with its objective principles of price formation, transform the game and the principles of price formation' (2004, p. 63). The new wave of open and peer production practices and processes responds to and drives a reorientation of the possibilities individuals have to participate directly in the production and management of knowledge. However, the direction this reorientation will take and the role of scientists' culture in this process is still to be deciphered. Open science advocates somehow foresee an open science renaissance enacted by scientists' willingness to share data and knowledge, new information and communication technologies, and a new system of incentives based on the acceptance of the open source model by both publicly funded research agencies and private corporations.

The stories I have presented here surely represent a shift towards more open approaches both in the public and private sector. They are rebellions and challenges to the incumbents of the current life sciences system – what I have called Big Bio to highlight the role of big corporations, global universities and international regulatory agencies. Yet they also show how this open science is strictly related to entrepreneurship, academic capitalism and neoliberalism. Its emphasis on cooperation, freedom from bureaucracies, openness and horizontality is reminiscent of capitalism's 'new spirits' described in several accounts of new information societies.

Open science and open source models acquire public value because they are an important part of the configuration of the relations between research, society and the market. This pushes us to place biohackers within the broader context of practices related to hacking that have assumed increasing importance in our societies. The spillover outside the software world that hacker cultures are experiencing is contaminating several sectors of culture and information production. Examples such as: the distributed hacker network Anonymous; Wikileaks and its emphasis on radical transparency and anti-bureaucratic claims; the *maker* movement, with its attempt at exporting open source software practices from bits to atoms; the hardware hacking and 3D printing movement; and the experience of The Pirate Bay and the emergence of Pirate Parties in Sweden and other North European countries; are testimony to a world in which transparency, sharing, distributed creativity, peer production, and distrust for institutions and bureaucracies are becoming more and more widespread.

All these examples are appearing in sectors that are strategic for contemporary capitalism: information production and management, political transparency and power over digital networks. Biohackers show that the roles that critiques of capitalism play are still of vital importance, as they provide ideologies that can be inscribed in the cultural context in which capitalism itself is developing, in order to be combined with its needs for accumulation. They drive and legitimise capitalism's transformations, while making new forms of accumulation attractive to people who still refer to older values and practices.

Capitalism needs its enemies for the sake of its own evolution, as it changes by mobilising critiques and opposing cultures and incorporating them into new cultural frameworks adapted to corporate goals. Studies of hacking have highlighted the constitutive complexity and the radical interplay of the different facets of hacker cultures, and their contribution to the evolution of life science research means that biological science is about to be infused with the same kind of contestations and contradictions that characterise software hacking.

This is due to the fact that hackers are not simply engaged in hands-on approaches to technology: their practices are a means of creating new politics. They are geared towards the development of concepts that lie at the very core of our societies, such as openness, property, freedom and autonomy. It is also for this very reason that studies of hacker cultures and practices are indispensable when trying to make sense of the evolution of science and technology in the information society at large.

Yet the history of computers shows that it is impossible to separate clearly the alternative and radical utopias of hackers from their contribution to the IT industry's development. In this sense, hacker cultures do not seem to be the object of capitalism's co-optation or absorption. Rather, they seem to have a constitutive role in the evolution of digital capitalism and thus of contemporary liberal societies. In spite of all their diversity and complexity, biohackers seem to have a strong agency and to be able to drive life sciences evolution, and thus interrogate us on the future of information societies.

Open science is a rich phenomenon, and in this book I did not intend to cover all of its complexity. For example, many open science projects and practices are driven by the commitment to free speech that characterises free software, especially in biomedical research. Others express the will to enact a more cooperative knowledge production within current science institutions. Yet biohackers are an important part of open science politics, as they represent two opposite and yet intertwined tendencies within this framework. One tendency lies towards an individualistic culture of openness both in information circulation and in capitalist competition, a new open frontier for science entrepreneurship in a new territory of accumulation. The other lies towards a collective, peer produced biology where open sharing is coupled with open participation and a discourse of democracy.

We cannot tell which one will eventually prevail, or in which way they will change the relationship between capital, biomedicine and society. Yet this ambivalence does not prevent them from posing a challenge to Big Bio, as shown by their radical opening to participation, their attempts at disrupting some of the concentrations of power that are typical of today's life sciences and their attacks against sites of appropriation that are hegemonic within the life sciences: intellectual property rights and restrictions to access. Therefore all the different directions that biohackers are taking are somehow part of a counter-symbolic order, since they challenge Big Bio's concentration of power as it exists today.

However, neither of these tendencies excludes a crucial role for entrepreneurship and profit. Genes and cells can always be objects of private interest. Biohackers are changing biocapitalism, intervening in the dynamics that hold the distribution of power over information and knowledge production. If we were to translate the claims that informed the free culture movement into biology, then 'genes want to be free' would be a slogan for biohacking. Yet throughout the history of the information society we have learnt that these claims are ambivalent. 'All information must be free!' can refer to free not as in 'free beer' or 'free speech', but as in 'free market'. This ambivalence now echoes again in labs and databases.

Notes

CHAPTER 1

1. I borrow this definition from Christopher Kelty (2010).
2. http://en.wikipedia.org/wiki/Open_source, accessed October 2012.
3. Other studies suggest that anticommons caused by gene patenting have never materialised, even though those fears continue to have an important effect on policymaking (Caulfield et al. 2006).
4. David (2003, p. 13); for an example of anticommons in genomics see Maurer (2006).
5. Heller and Eisenberg (1998); see also Nowotny et al. (2001); Hedgecoe and Martin 2008. For a historical perspective on open science see David (2003) and Eamon 1990.
6. Castells (1996); Coleman and Golub (2008); Himanen and Torvalds (2001); on this ambivalence that I will further analyse in Chapters 2 and 3, see Biagioli (2006); Castells (1996 and 2005); Coleman (2004); Coleman and Golub (2008); Johns (2009); Mattelart (2003).
7. Other works on hacker ethic, and other versions of it, are Best (2003b); Ippolita (2005); Jesiek (2003); Moody (2001); Raymond (2001).

CHAPTER 2

1. A project launched by Lawrence Lessig, Creative Commons writes licenses alternative to copyright. CC licenses typically allow people to copy, share and even modify cultural products such as texts, music or images. www.creative-commons.org, accessed October 2012.
2. http://publicaccess.nih.gov/, accessed January 2012.
3. www.telethon.it/en/scientists/open-access, accessed October 2012.
4. www.ornl.gov/sci/techresources/Human_Genome/research/bermuda.shtml, accessed October 2012.
5. For a well-known example of open innovation outside the domain of scientific research see Huston and Sakkab (2006).
6. www.arduino.cc, accessed June 2012.
7. http://pinkarmy.org/about/, accessed December 2012.
8. Talk at the Open Science Summit, Berkeley, 30 July 2010.
9. Halloween I. http://catb.org/~esr/halloween/halloween1.html, accessed January 2011.
10. Refer to Bauwens (2005); Benkler (2006). For a study specifically focused on open biology see Hope (2008).
11. The traditional ceremony in which indigenous peoples of the Pacific Northwest Coast redistribute wealth. Potlatch means 'to give away' or 'a gift'. According to Wikipedia, 'The status of any given family is raised not by who has the most resources, but by who distributes the most resources', http://en.wikipedia.org/wiki/Potlatch, accessed January 2011.

12. See for example Paul Rabinow's 'biosociality' (1996) and Sjeila Jasanoff's 'technologies of humility' (2003).

CHAPTER 3

1. For an explicit comparison between the hacker ethic, free software and open science see Kelty (2001) and Willinsky (2005).
2. Adapted from Levy (2010a, pp. 27–38). Other overviews of hacker values overlap Levy's one. In *Rebel code*, a history of the free software and open source movements, Moody uses terms such as openness, sharing, cooperation, freedom, community, creation, beauty and joy (2001).
3. Quoted in Levy (2010a, p. 180).
4. O'Neil (2009); see also Castells (2005) and Coleman (2004). Chapter 2 includes a discussion of the role of participation and sharing in current digital capitalism.
5. www.hackmeeting.org, accessed October 2012.

CHAPTER 4

1. www.sorcerer2expedition.org, accessed January 2010.
2. http://camera.calit2.net, accessed January 2010.
3. http://en.wikipedia.org/wiki/Metagenomics, accessed October 2012.
4. http://partsregistry.org/Main_Page, accessed October 2012.
5. Press conference of 4 March 2004, quoted in Pollack (2007).
6. On secrecy, see Louis et ai. (2001); Blumenthal et al. (1996); on anticommons, Heller and Eisenberg (1998); on the problems posed by the use of information contained in databases for the advancement of science, see Gardner and Rosenbaum (1998).
7. www.sagebase.org/commons/, accessed November 2012.
8. The scientific institutions, public and private, appearing in the scientific articles published by *Science* and *PLoS Biology* are: J. Craig Venter Institute (JCVI); California Institute for Telecommunications and Information Technology (Calit2); University of California San Diego (UCSD); University of California Irvine; UCSD Center for Earth Observations and Applications; San Diego Supercomputer Center; University of California Davis; Department of Biological Sciences, University of Southern California; Your Genome, Your World; Departmento de Ecología Evolutiva, Instituto de Ecología, Universidad Nacional Autónoma de México; Department of Oceanography, University of Hawaii; Bedford Institute of Oceanography; Smithsonian Tropical Research Institute, Panama; Departamento de Oceanografía, Universidad de Concepción, Chile; Universidad de Costa Rica; Department of Environmental Sciences, Rutgers University; Department of Earth Sciences, University of Southern California; Razavi-Newman Center for Bioinformatics, Salk Institute for Biological Studies; Burnham Institute for Medical Research; University of California Berkeley; Physical Biosciences Division, Lawrence Berkeley National Laboratory; Brown University.
9. For example, the sequencing and the publication of his own genome, which inspired him to write his autobiography (Venter 2007b), or else the production of an artificial microbial genome (Gibson et al. 2008).

CHAPTER 5

1. www.gisaid.org, accessed February 2011.
2. Acronym for Differentiating Infected Animals from Vaccinated Animals.
3. A network of expertise on animal influenzas, www.offlu.net/index.html, accessed February 2011.
4. Respectively: Food and Agriculture Organization and World Organisation for Animal Health.
5. http://grants.nih.gov/grants/policy/data_sharing/, accessed February 2011.
6. www.ornl.gov/sci/techresources/Human_Genome/research/bermuda.shtml, accessed February 2011.
7. A 'global electronic reporting system for outbreaks of emerging infectious diseases', www.promedmail.org, accessed September 2010.
8. World Health Organization, document EB120/INF.DOC./3, 11 January 2007, p. 2.
9. See the discussion in Chapter 2 around Barbara McKlintock's case, (Keller 1983).
10. Interview with Ilaria Capua, January 2011.
11. Ibid.
12. www.gisaid.org, accessed February 2011.
13. Ibid.
14. Cited in Johns (2009a).
15. GISAID EpiFlu Database access agreement, www.gisaid.org, accessed February 2011.
16. Interview with Ilaria Capua, January 2011.
17. Ibid.
18. GISAID EpiFlu Database access agreement.

CHAPTER 6

1. www.diybio.org, accessed October 2012; for a clear and comprehensive history of DIYbio's birth refer to Wohlsen (2011).
2. See for example Bloom (2009); Kelty (2010); Ledford (201); Schmidt (2008).
3. See Critical Art Ensemble (2002).
4. http://diybio.org/local, accessed October 2012.
5. DIYbio FAQ, http://openwetware.org/wiki/DIYbio/FAQ, accessed January 2011.
6. www.codecon.org/2009/program.html, accessed January 2011.
7. Interviewee A, San Francisco, July 2010.
8. Obviously, this is somehow part of the hacker mythology: Steve Wozniak, for example, made it clear that Apple computers were not designed and built in Jobs' garage.
9. Interviewee B, San Francisco, July 2010.
10. Interviewee A, San Francisco, July 2010.
11. Interviewee B, San Francisco, July 2010.
12. DIYbio FAQ.
13. Interviewee C, Los Angeles, October 2010.
14. www.igem.org, accessed October 2012. See also Alper (2009).
15. http://artisopensource.net/cure, accessed October 2012.

16. Interviewee D, Seattle, July 2010.
17. Interviewee B, San Francisco, July 2010.
18. Interviewee C, Los Angeles, October 2010.
19. Interviewee E, Los Angeles, November 2010.
20. Interviewee F, San Francisco, October 2010.
21. Interviewee G, San Francisco, October 2010.
22. Interviewee B, San Francisco, July 2010.

Bibliography

Aime, M. and Cossetta, A., 2010. *Il dono al tempo di Internet*. Torino: Einaudi.

Alper, J., 2009. 'Biotech in the basement', *Nature Biotechnology* 27 (12), pp. 1077–8.

Altman, M., 2012. 'Hacking at the crossroads: US military funding of hackerspaces', *Journal of Peer Production* 3 (2). http://peerproduction.net/issues/issue-2/invited-comments/hacking-at-the-crossroad/, accessed October 2012.

American Coalition Against Biopiracy, 2006. *Captain Hook awards for biopiracy*. http://www.captainhookawards.org/winners/2006_pirates, accessed August 2008.

Anonymous, 2006a. 'Secret avian flu archive'. *The New York Times*, 15 March. http://www.nytimes.com/2006/03/15/opinion/15wed4.html?_r=0, accessed October 2012.

Anonymous, 2006b. 'Dreams of flu data', *Nature*, 440 (7082), pp. 255–6.

Anonymous, 2008. 'The game changers'. *Seed*, December, pp. 80–4.

Anonymous, 2010. 'Learning to share'. *Nature* 463 (7280), p. 401.

Arvidsson, A. and Colleoni, E., 2012. 'Value in informational capitalism and on the internet', *The Information Society*, 28 (3), pp. 135–50.

Barbrook, R., 1998. 'The hi-tech gift economy', *First Monday*, 3 (12). http://firstmonday.org/htbin/cgiwrap/bin/ojs/index.php/fm/article/view/631/552, accessed September 2010.

Barbrook, R. and Cameron, A., 1996. 'The Californian ideology', *Science as Culture*, 6 (1), pp. 44–72.

Bauwens, M., 2005. 'The political economy of peer production', *CTheory*. http://www.ctheory.net/articles.aspx?id=499+, accessed January 2011.

Bauwens, M., 2010. 'Is there something like a peer-to-peer science?', *JCOM* 9 (1). jcom.sissa.it/archive/09/01/Jcom0901(2010)C01/Jcom0901(2010)C02, accessed January 2011.

Bauwens, M., et al., 2012. *Report: a synthetic overview of the collaborative economy*. Orange labs and P2P Foundation. http://p2pfoundation.net/Synthetic_Overview_of_the_Collaborative_Economy, accessed October 2012.

Bell, D., 1973. *The coming of post-industrial society*. New York: Basic Books.

Benkler, Y., 2006. *The wealth of networks*. New Haven: Yale University Press.

Best, K., 2003a. 'Beating them at their own game', *International Journal of Cultural Studies* 6 (4), pp. 449–70.

Best, K., 2003b. 'The hackers challenge: active access to information, visceral democracy and discursive practice', *Social Semiotics* 13 (3), pp. 263–82.

Biagioli, M., 2006. 'Patent republic: representing inventions, constructing rights and authors', *Social Research* 73 (4), pp. 1129–72.

Björk B.C., et al., 2010. 'Open access to the scientific journal literature: situation 2009', *PLoS ONE* 5 (6). http://www.plosone.org/article/info:doi/10.1371/journal.pone.0011273, accessed October 2012.

Bloom, J., 2009. 'The geneticist in the garage', *The Guardian*, 19 March. http://www.guardian.co.uk/technology/2009/mar/19/biohacking-genetics-research, accessed October 2012.

Blumenthal, D., et al., 1996. 'Withholding research results in academic life science: evidence from a national survey of faculty', *Journal of the American Medical Association* 277, pp. 1224–8.

Bobe, J., 2008. 'Science without scientists'. http://diybio.org/blog/science-without-scientists, accessed September 2010.

Bogner, P., et al., 2006. 'A global initiative on sharing avian flu data', *Nature* 442, p. 981.

Bok, S., 1982. 'Secrecy and openness in science: ethical considerations', *Science, Technology and Human Values* 7 (38), pp. 32–41.

Boltanski, L. and Chiappello, E., 2005. *The new spirit of capitalism*. New York: Verso.

Boltanski, L. and Thévenot, L., 1999. 'The sociology of critical capacity', *European Journal of Social Theory* 2 (3), pp. 359–77.

Bourdieu, P., 2004. *Science of science and reflexivity*. Cambridge: Polity Press.

Boustead, G., 2008. 'The biohacking hobbyist', *Seed Magazine*. http://seedmagazine.com/content/article/the_biohacking_hobbyist/, accessed January 2011.

Brewer, J. and Staves, S., 1996. *Early modern conceptions of property*. London and New York: Routledge.

Bucchi, M., 1998. *Science and the media: alternative routes in scientific communication*. London and New York: Routledge.

Budapest Open Access Initiative, 2001. http://www.soros.org/openaccess/read.shtml, accessed September 2010.

Bush, V., 1945. *Science, the endless frontier*. http://www.nsf.gov/od/lpa/nsf50/vbush1945.htm, accessed January 2011.

Butler, D., 2005. 'Flu researchers slam US agency for hoarding data', *Nature* 437, pp. 458–9.

Calabrese, F., 2006. 'Agraria, ricerca vicina ai produttori siciliani', *La Repubblica*, 25 March, p. 16.

Callon, M. and Rabeharisoa, V., 2003. 'Research "in the wild" and the shaping of new social identities', *Technology in Society* 25, pp. 193–204.

Calvert, J., 2010. 'Synthetic biology: constructing nature?', *The Sociological Review* 58 (s1), pp. 95–112.

Capua, I., et al., 2006. 'Veterinary virologists share avian flu data', *Science* 312 (5780), p. 1597b.

Capua, I., 2009a. Speech at the Venice sessions conference, 31 March. http://www.youtube.com/watch?v=R1UqY_4VLI0, accessed January 2011.

Capua, I., 2009b. 'Ilaria Capua: la scienza open source', *Wired Italia*. http://mag.wired.it/rivista/storie/ilaria-capua-la-scienza-open-source.html, accessed January 2011.

Carlson, R., 2005. 'Splice it yourself', *Wired*. http://www.wired.com/wired/archive/13.05/view.html?pg=2, accessed January 2011.

Castelfranchi, Y., 2004. 'When the data isn't there. Disclosure: the scientific community (and society) at a crossroads', *JCOM* 3 (2). http://jcom.sissa.it/archive/03/02/F030201, accessed September 2010.

Castells, M., 1996. *The rise of the network society*. Oxford: Blackwell.

Castells, M., 2005. *Open source as social organization of production and as a form of technological innovation based on a new conception of property rights*, presentation at the World Social Forum. http://www.choike.org/nuevo_eng/informes/2623.html, accessed January 2011.

Caulfield, T., et al., 2006. 'Evidence and anecdotes: an analysis of human gene patenting controversies', *Nature Biotechnology* 24 (9), pp. 1091–4.

Cavadini, F., 2006. 'Aviaria, la scienziata italiana si ribella: "le mie scoperte? Sul web, per tutti"', *Corriere della Sera*, 14 March, p. 21.

Church, G., et al., 2012. 'Next-generation digital information storage in DNA', *Science* 337 (6102), p. 1628.

Coleman, G., 2004. 'The political agnosticism of free and open source software and the inadvertent politics of contrast', *Anthropological Quarterly* 77 (3), pp. 507–19.

Coleman, G. and Golub, A., 2008. 'Hacker practice: moral genres and the cultural articulation of liberalism', *Anthropological Theory* 8 (3), pp. 255–77.

Collins, H. and Evans, R., 2007. *Rethinking expertise*. Chicago: Chicago University Press.

Collins, H. and Pinch, T., 1994. *The golem: what everyone should know about science*. Cambridge: Cambridge University Press.

Comfort, N., 2008. 'Rebellion and iconoclasm in the life and science of Barbara McClintock', in Harman, O. and Dietrich, M., *Rebels, mavericks, and heretics in biology*. New Haven: Yale University Press, pp. 137–53.

Conner, C., 2005. *A people's history of science: miners, midwives, and low mechanics*. New York: Nation Books.

Conover, D., 2005. 'Cracking the ocean code' [DVD]. *Discovery Channel*.

Cooper, M., 2008. *Life as surplus: biotechnology and capitalism in the neoliberal era*. Seattle and London: University of Washington Press.

Coyaud, S., 2007. 'Una diva contro il virus', *La Repubblica delle Donne*, 2 February, pp. 77–8.

Coyaud, S., 2008a. 'Influenza aviaria: capua vince', *La Repubblica delle Donne*, 20 September, p. 66.

Coyaud, S., 2008b. 'I sogni di una diva', *La Repubblica delle Donne*, 8 November, p. 42.

Critical Art Ensemble, 2002. *The molecular invasion*. New York: Autonomedia.

Dányi, E., 2006. 'Xerox project: photocopy machines as a metaphor for an "open society"', *The Information Society* 22, pp. 111–15.

Darwin, C., 1958. *The autobiography of Charles Darwin, 1809–1882: with original omissions restored*. Edinburgh: Collins.

David, P., 2001. *From keeping 'nature's secrets' to the institutionalization of 'open science'*, Oxford University Economic and Social History Series 023, Economics Group, University of Oxford.

David, P., 2003. *The economic logic of 'open science' and the balance between private property rights and the public domain in scientific data and information: a primer*, Stanford Institute for Economic Policy Research, discussion paper.

Davies, K., 2002. *Cracking the genome: inside the race to unlock human DNA*. Baltimore: Johns Hopkins University Press.

Deleuze, G., 1992. 'Postscript on the societies of control', *October* 59, pp. 3–7.

Delfanti, A. (ed.), 2010. 'Users and peers: from citizen science to P2P science', *JCOM* 9 (1). jcom.sissa.it/archive/09/01/Jcom0901(2010)E/, accessed January 2011.

Delfanti, A., 2011. 'Hacking genomes: the ethics of open and rebel biology', *International Review of Information Ethics* 15, pp. 52–7.

Delfanti, A., 2012. 'Tweaking genes in your garage: biohacking between activism and entrepreneurship', in Sützl, W. and Hug, T., *Activist media and biopolitics.* Innsbruck: Innsbruck University Press, pp. 163–76.

Delfanti, A., Castelfranchi, Y. and Pitrelli, N., 2009. 'What Dr. Venter did on his holidays: exploration, hacking, entrepreneurship in the narratives of the Sorcerer II expedition', *New Genetics and Society,* 28 (4), pp. 415–30.

Despret, V., 2008. 'Culture and gender do not dissolve into how scientists "read" nature: Thelma Rowell's heterodoxy', in Harman, O. and Dietrich, M., *Rebels, mavericks, and heretics in biology.* New Haven: Yale University Press.

Discovery Channel, 2005. 'About discovery channel quest', *Discovery channel.* http:// dsc.discovery.com/convergence/quest/about-quest.html, accessed August 2008.

DIYbio, 2010. Draft letter to the PCSBI commission, personal email from Jason Bobe.

Dutfield, G., 2003. *Intellectual property rights and the life sciences industries: a twentieth century history.* Aldershot: Ashgate.

Dyson, F., 2006. *The scientist as rebel.* New York: New York Review of Books.

Dyson, F., 2009. 'When science & poetry were friends', *The New York review of books,* 13 August.

Eamon, W., 1990. 'From the secrets of nature to public knowledge', in Lindberg, D. and Westman, R., (eds) *Reappraisals of the scientific revolution.* Cambridge: Cambridge University Press, pp. 333–66.

Economist, 2007. 'Sorcerer's apprenticeship: what Dr Venter did on his holidays', *The Economist,* 15 March.

Economist, 2009. 'Hacking goes squishy', *The Economist,* 5 September, pp. 30–1.

Eisenberg, R., 2006. 'Patents and data-sharing in public science', *Industrial and Corporate Change* 15 (6), pp. 1013–31.

Enserink, M., 2006a. 'As H5N1 keeps spreading, a call to release more data', *Science* 311 (5765), p. 1224.

Enserink, M., 2006b. 'Italy's influenza diva', *Science* 314 (5801), pp. 918–19.

ETC Group, 2004. *Communiqué 84: playing God in the Galapagos.* http://www. etcgroup.org/_page24?id=120, accessed August 2008.

Feuer, L., 1963. *The scientific intellectual: the psychological and sociological origins of modern science.* New York: Basic Books.

Fortun, M., 2001. 'Mediated speculations in the genomics futures markets', *New Genetics and Society* 20 (2), pp. 139–56.

Franklin, S. and Lock, M., 2003. 'Animation and cessation', in Franklin, S. and Lock, M., (eds) *Remaking life and death: toward an anthropology of the biosciences.* Santa Fe: SAR Press, pp. 3–22.

Fuchs, S., 1993. 'Positivism is the organizational myth of science', *Perspectives on Science* 1, pp. 1–23.

Gardner, W. and Rosenbaum, J., 1998. 'Database protection and access to information', *Science* 281 (5378), pp. 786–7.

Gates, B., 1976. 'An open letter to hobbyists', *Homebrew Computer Club Newsletter* 2 (1), p. 2.

Gibbons, M., et al., 1994. *The new production of knowledge: the dynamics of science and research in contemporary societies.* London: Sage.

Gibson, D., et al., 2008. 'Complete chemical synthesis, assembly, and cloning of a mycoplasma genitalium genome', *Science* 319 (5867), pp. 1215–20.

Gibson, D., et al., 2010. 'Creation of a bacterial cell controlled by a chemically synthesized genome', *Science* 329 (5987), pp. 52–6.

Golob, J., 2007. 'Homebrew molecular biology club', *The Stranger*. http://slog. thestranger.com/2007/11/homebrew_molecular_biology_club, accessed January 2011.

Goodell, R., 1977. *The visible scientists*. London: Little, Brown.

Gross L., 2007. 'Untapped bounty: sampling the seas to survey microbial biodiversity', *PLoS Biology* 5 (3). http://www.plosbiology.org/article/info%3Adoi%2F10.1371%2Fjournal.pbio.0050085, accessed October 2012.

Gruppo Laser, 2005. *Il sapere liberato*. Milano: Feltrinelli.

Guthrie, J., 2009. 'Do-it-yourself biology on rise', *San Francisco Chronicle*, 20 December.

Hackett, E., 1990. 'Science as a vocation in the 1900s', *Journal of Higher Education* 61 (3), pp. 241–79.

Hagstrom, W.O., 1982. 'Gift giving as an organisational principle in science', in Barnes, B. and Edge, D. (eds) *Science in context: readings in the sociology of science*. Cambridge: MIT Press.

Haraway, D., 1988. 'The biopolitics of postmodern bodies: constitutions of self in immune system discourse', *Differences* 1 (1), pp. 3–43.

Haraway, D., 1991. 'A cyborg manifesto: science, technology, and socialist-feminism in the late twentieth century', in Haraway D. *Simians, cyborgs and women: the reinvention of nature*. New York: Routledge, pp. 149–81.

Hardin, G., 1968. 'The tragedy of the commons', *Science* 162 (3859), pp. 1243–8.

Hardt, M. and Negri, A., 2000. *Empire*. Cambridge: Harvard University Press.

Harman, O. and Dietrich, M., 2008. *Rebels, mavericks, and heretics in biology*. New Haven: Yale University Press.

Harvey, D., 2005. *A brief history of neoliberalism*. Oxford: Oxford University Press.

Hayles, K., 1999. *How we became posthuman: virtual bodies in cybernetics, literature, and informatics*. Chicago: University of Chicago Press.

Hedgecoe, A. and Martin, P., 2008. 'Genomics, STS, and the making of sociotechnical futures', in Hackett, E., et al. (eds) *The handbook of science and technology studies*. Boston: MIT Press, pp. 817–39.

Heller, M. and Eisenberg, R., 1998. 'Can patents deter innovation? The anticommons in biomedical research', *Science* 280 (5364), pp. 698–701.

Helmreich, S., 2007. 'Blue-green capital, biotechnological circulation and an oceanic imaginary: a critique of biopolitical economy', *Biosocieties* 2, pp. 287–302.

Hilgartner, S., 1995. 'Biomolecular databases: new communication regimes for biology?' *Science Communication* 17 (2), pp. 240–63.

Hilgartner, S. and Brandt-Rauf, S., 1994. 'Data access, ownership, and control', *Science Communication* 11 (4), pp. 355–72.

Himanen, P. and Torvalds, L., 2001. *The hacker ethic, and the spirit of the information age*. New York: Random House.

Hope, J., 2008. *Biobazaar: the open source revolution and biotechnology*. Cambridge: Harvard University Press.

Hough, A. and Bingham, J., 2010. 'WikiLeaks: Julian Assange supporters must raise £200,000 bail in cash after credit card crisis', *The Telegraph*, 15 December.

Huston, L. and Sakkab, N., 2006. 'Connect and develop: inside Procter & Gamble's new model for innovation', *Harvard Business Review* 84, pp. 58–66.

Ippolita, 2005. *Open non è free. Comunità digitali tra etica hacker e mercato globale*. Milano: Eleuthera.

Jasanoff, S., 2003. 'Technologies of humility: citizen participation in governing science', *Minerva* 41 (3), pp. 223–44.

JCVI (J. Craig Venter Institute), 2004a. *Global ocean sampling expedition: fact sheet – expedition overview.* http://www.jcvi.org/cms/fileadmin/site/research/projects/gos/Expedition_Overview.pdf, accessed August 2008.

JCVI (J. Craig Venter Institute), 2004b. *IBEA researchers publish results from environmental shotgun sequencing of sargasso sea.* http://www.jcvi.org/cms/press/press-releases/full-text/browse/8/article/ibea-researchers-publish-results-from-environmental-shotgun-sequencing-of-sargasso-sea-in-science-d/?tx_ttnews[backPid]=67&cHash=22fe240eeb, accessed August 2008.

JCVI (J. Craig Venter Institute), 2006. *About the J. Craig Venter Institute.* http://www.jcvi.org/cms/about/overview/, accessed August 2008.

JCVI (J. Craig Venter Institute), 2007. *Minimal bacterial genome.* United States patent application 20070122826, 31 May.

JCVI (J. Craig Venter Institute), 2010. *First self-representation on synthetic bacterial cell.* http://www.jcvi.org/cms/press/press-releases/full-text/article/first-self-replicating-synthetic-bacterial-cell-constructed-by-j-craig-venter-institute-researcher/, accessed January 2011.

Jenkins, H., 2008. *Convergence culture: where old and new media collide.* New York: New York University Press.

Jesiek, B., 2003. 'Democratizing software: open source, the hacker ethic, and beyond', *First Monday* 8 (10). http://firstmonday.org/htbin/cgiwrap/bin/ojs/index.php/fm/article/view/1082, accessed October 2012.

Johns, A., 2009a. *Piracy.* Chicago: University of Chicago Press.

Johns, A., 2009b. 'Piracy as a business force', *Culture Machine* 10, pp. 44–63. http://culturemachine.net/index.php/cm/article/viewArticle/345, accessed June 2012.

Johnson C., 2008. 'Accessible science: hackers aim to make biology household practice', *The Boston Globe*, 15 September.

Kay, L., 2000. *Who wrote the book of life? A history of the genetic code.* Stanford: Stanford University Press.

Keller, E.F., 1983. *A feeling for the organism: the life and work of Barbara McClintock.* New York: Freeman.

Keller, E.F., 2000. *The century of the gene.* Cambridge: Harvard University Press.

Keller, E.F., 2009. 'What does synthetic biology have to do with biology?' *Biosocieties* 4 (2–3), pp. 291–302.

Kelty, C., 2001. 'Free software/free science', *First Monday* 6 (12). http://firstmonday.org/htbin/cgiwrap/bin/ojs/index.php/fm/article/view/902/811, accessed June 2012.

Kelty, C., 2008. *Two bits: the cultural significance of free software.* Durham: Duke University Press.

Kelty, C., 2010. 'Outlaw, hackers, Victorian amateurs: diagnosing public participation in the life sciences today', *JCOM* 9 (1). http://jcom.sissa.it/archive/09/01/Jcom0901(2010)C01/Jcom0901(2010)C03, accessed June 2012.

Kevles, D., 2011. 'New blood, new fruits', in Biagioli, M., et al. (eds) *Making and unmaking intellectual property.* Chicago: University of Chicago Press, pp. 253–68.

Kohler, R., 1994. *Lords of the fly: Drosophila genetics and the experimental life.* Chicago: Chicago University Press.

Krimsky, S., 2006. 'Autonomy, disinterest, and entrepreneurial science', *Society* 43 (4), pp. 22–9.

Kropotkin, P., 1996 (1901). *Fields, factories and workshops*. Montreal: Black Rose Books.

Kuhn, T., 1996 (1962). *The structure of scientific revolutions*. Chicago: University of Chicago Press.

Kumar, K., 2004. *From post-industrial to post-modern dociety: new theories of the contemporary world*. Hoboken: Wiley-Blackwell.

Lam, A., 2010. 'From "ivory tower traditionalists" to "entrepreneurial scientists"?: academic scientists in fuzzy university–industry boundaries', *Social Studies of Science* 40 (2), pp. 307–40.

Laudan, L., 1982. 'Two puzzles about science: reflections on some crises in the philosophy and sociology of science', *Minerva* 20 (3–4) pp. 253–68.

Ledford, H., 2010. 'Life hackers', *Nature* 467, pp. 650–2.

Levina, M., 2010. 'Googling your genes: personal genomics and the discourse of citizen bioscience in the network age', *JCOM* 9 (1). http://jcom.sissa.it/archive/09/01/Jcom0901(2010)A06/, accessed October 2012.

Levy, S., 2010a (1984). *Hackers: heroes of the computer revolution*. Cambridge: O'Reilly.

Levy, S., 2010b. 'Geek power: Steven Levy revisits tech titans, hackers, idealists', *Wired*. http://www.wired.com/magazine/2010/04/ff_hackers/all/1, accessed January 2012.

Lewenstein, B., 1995. 'From fax to facts: communication in the cold fusion saga', *Social Studies of Science* 25 (3), pp. 403–36.

Lewontin, R., 2008. 'Legitimation is the name of the game', in Harman, O. and Dietrich, M., *Rebels, mavericks, and heretics in biology*. New Haven: Yale University Press, pp. 372–83.

Leydesdorff, L. and Meyer, M., 2010. 'The decline of university patenting and the end of the Bayh–Dole effect', *Scientometrics* 83 (2), pp. 355–62.

Liakopoulos, M., 2002. 'Pandora's box or panacea? Using metaphors to create the public representations of biotechnology', *Public Understanding of Science* 11 (1), pp. 5–32.

Louis, K.S., et al., 2001. 'Entrepreneurship, secrecy and productivity: a comparison of clinical and non-clinical faculty', *Journal of Technology Transfer* 26 (3), pp. 233–45.

Luther, M., 1517. *Ninety-five theses on the power and efficacy of indulgences*. http://en.wikisource.org/wiki/The_Ninety-Five_Theses, accessed December 2012.

Martin, P., et al., 2010. 'Genomics, the crisis of pharmaceutical productivity and the search for sustainability', in Atkinson, P., et al (ed.) *Handbook of genetics and society*. London: Routledge.

Marx, K., 1990 (1867). *Capital volume I*. New York: Penguin.

Mattelart, A., 2003. *The information society: an introduction*. London: Sage Publications.

Maurer, S., 2006. 'Inside the anticommons: academic scientists' struggle to build a commercial self-supporting human mutations database, 1999–2001', *Research Policy* 35 (6), pp. 839–53.

Mauss, M., 2002 (1922). *The gift: the form and reason for exchange in archaic societies*. New York: W.W. Norton & Company.

Maxigas, 2012. 'Hacklabs and hackerspaces: genealogies', *Journal of Peer Production* 3 (2). http://peerproduction.net/issues/issue-2/peer-reviewed-papers/hacklabs-and-hackerspaces/, accessed October 2012.

McGuire, A., et al., 2010. 'Regulating direct-to-consumer personal genome testing', *Science* 330 (6001), pp. 181–2 .

The Mentor, 1986. 'The hacker manifesto', *Phrack* 1 (7). http://www.phrack.org/issues.html?issue=7&id=3&mode=txt, accessed December 2012.

Merton, R., 1973 (1942). 'The normative structure of science', in Merton, R., *The sociology of science: theoretical and empirical investigations*. Chicago: University of Chicago Press, pp. 221–78.

Meyer, M., 2012. 'Build your own lab: do-it-yourself biology and the rise of citizen biotech-economies', *Journal of Peer Production* 2. http://peerproduction.net/issues/issue-2/invited-comments/build-your-own-lab/, accessed October 2012.

Mikkonen, T., Vadén, T. and Vainio, N., 2007. 'The Protestant ethic strikes back: open source developers and the ethic of capitalism', *First Monday* 12 (2). http://firstmonday.org/htbin/cgiwrap/bin/ojs/index.php/fm/article/viewArticle/1623/1538, accessed January 2012.

Mills, E. and Tereskerz, P., 2007. 'Changing patent strategies: what will they mean for the industry?', *Nature Biotechnology* 25 (8), pp. 867–68.

Mirowski, P. and Sent, E., 2008. 'The commercialization of science and the response of STS', in Hackett, E., et al. (eds) *The handbook of science and technology studies*. Cambridge: MIT Press, pp. 635–90.

Mitroff, I.I., 1974. *The subjective side of science*. Amsterdam, New York: Elsevier.

Moody, G., 2001. *Rebel code: the inside story of Linux and the open source*. Cambridge: Perseus.

Morange, M., 2008. 'Rebels? No, simply scientists', *PloS Biology* 6 (9). http://www.plosbiology.org/article/info%3Adoi%2F10.1371%2Fjournal.pbio.0060242, accessed October 2012.

Murray, F., 2011. 'Patenting life: how the Oncomouse patent changed the lives of mice and men', in Biagioli, M., et al., (eds) *Making and unmaking intellectual property*. Chicago: University of Chicago Press, pp. 399–412.

Nair, P., 2009. 'Straight talk with ... Mac Cowell and Jason Bobe', *Nature Medicine* 15 (3), pp. 230–1.

Nerlich, B. and Hellsten, I., 2004. 'Genomics: shift in metaphorical landscape between 2000 and 2003', *New Genetics and Society* 23 (3), pp. 255–68.

Nicholls, H., 2007. 'Sorcerer II: the search for microbial diversity roils the waters', *PLoS Biology*, 5 (3). http://www.plosbiology.org/article/info%3Adoi%2F10.1371%2Fjournal.pbio.0050074, accessed October 2012.

Nielsen, M., 2012. *Reinventing discovery: the new era of networked science*. Princeton: Princeton University Press.

Nowotny, H., 1993. 'Socially distributed knowledge: five spaces for science to meet the public', *Public Understanding of Science* 2, pp. 307–19.

Nowotny, H., Scott, P. and Gibbons, M., 2001. *Re-thinking science: knowledge production in an age of Uncertainty*. Cambridge: Polity Press.

NPR, 2009. 'Taking biological research out of the laboratory', 27 December. http://www.npr.org/templates/story/story.php?storyId=121954328, accessed March 2010.

O'Neil, M., 2009. *Cyberchiefs: autonomy and authority in online tribes*. London: Pluto Press.

Oriani, R., 2006. 'Perché ho detto no al galateo della scienza', *Io Donna*, 6 May, pp. 71–4.

Ostrom, E., 1990. *Governing the commons: the evolution of institutions for collective action*. Cambridge: Cambridge University Press.

Packer, K. and Webster, A., 1996. 'Patenting culture in science: reinventing the scientific wheel of credibility', *Science, Technology and Human Values* 21 (4), pp. 427–53.

Panzieri, R., 1976. *Lotte operaie nello sviluppo capitalistico*. Torino: Einaudi.

Patterson, M., 2010. *A biopunk manifesto*. http://maradydd.livejournal.com/496085. html, accessed January 2011.

Pearson, H., 2006. 'Bird flu data liberated', *Nature News*. http://www.nature.com/news/2006/060821/full/news060821–10.html, accessed January 2011.

Pestre, D., 2008. 'Challenges for the democratic management of technoscience: governance, participation and the political today', *Science as Culture* 17 (2), pp. 101–119.

Pistoi, S., 2006. 'Liberate i dati su H5N1', *Le Scienze*, April, p. 31.

Polanyi, M., 1962. 'The republic of science', *Minerva* 1 (1), pp. 54–73. http://www.missouriwestern.edu/orgs/polanyi/mp-repsc.htm, accessed October 2012.

Pollack A., 2007. 'Groundbreaking gene scientist is taking his craft to the oceans', *New York Times*, 5 March. http://www.nytimes.com/2004/03/05/science/05GENE.html, accessed December 2012.

Popp Berman, E., 2008. 'Why did universities start patenting?', *Social Studies of Science* 38 (6), pp. 835–71.

Pottage, A., 2006. 'Too much ownership: bio-prospecting in the age of synthetic biology', *BioSocieties* 1, pp. 137–58.

Processed World, 1982. Issue 4. http://archive.org/stream/processedworld04proc#page/n1/mode/2up, accessed December 2012.

Rabinow, P., 1996. 'Artificiality and enlightenment: from sociobiology to biosociality', in Rabinow, P., *Essays on the anthropology of reason*. Princeton: Princeton University Press.

Rabinow, P., 1999. *French DNA: trouble in purgatory*. Chicago and London: University of Chicago Press.

Rai, A. and Boyle, J., 2007. 'Synthetic biology: caught between property rights, the public domain, and the commons', *PLoS Biology* 5 (3). http://www.plosbiology.org/article/info:doi/10.1371/journal.pbio.0050058, accessed October 2012.

Raymond, E., 2001. *The cathedral and the bazaar*. Cambridge: O'Reilly.

Rubin, J. and Cleaver, E., 1970. *Do it: scenarios of the revolution*. New York: Simon and Schuster.

Salzberg, S., Ghedin, E. and Spiro, D., 2006. 'Shared data are key to beating threat from avian flu', *Nature* 440, p. 605.

Schmidt, M., 2008. 'Diffusion of synthetic biology: a challenge to biosafety', *Systems and Synthetic Biology* 2, pp. 1–6.

Segerstrale, U., 2008. 'Against the grain: the science and life of William D. Hamilton', in Harman, O. and Dietrich, M., (eds) *Rebels, mavericks, and heretics in biology*. New Haven: Yale University Press, pp. 282–301.

Shapin, S., 2008. *The scientific life: a moral history of a late modern vocation*. Chicago: Chicago University Press.

Shiva, V., 1999. *Biopiracy: the plunder of nature and knowledge*. Cambridge: South End Press.

Shreeve, J., 2004a. *The genome war*. New York: Random House.

Shreeve, J., 2004b. 'Craig Venter's epic voyage to redefine the origin of the species', *Wired*, 12 (8). http://www.wired.com/wired/archive/12.08/venter.html, accessed October 2012.

Silber, I., 2003. 'Pragmatic sociology as cultural sociology: beyond repertoire theory?', *European Journal of Social Theory* 6 (4), pp. 427–49.

Simmons, M., 2007. 'Terrorizing the artists in the USA', *The Huffington Post*. http://www.huffingtonpost.com/michael-simmons/terrorizing-the-artists-i_b_52261.html, accessed January 2011.

Stodden, V., 2010a. 'Open science: policy implications for the evolving phenomenon of user-led scientific innovation', *JCOM* 9 (1). http://jcom.sissa.it/archive/09/01/Jcom0901(2010)A05/, accessed October 2012.

Stodden, V., 2010b. 'Two ideas for open science', talk given at the Open Science Summit, Berkeley, 29 July.

Suber, P., 2012. *Open access*. Cambridge: MIT Press.

Sunder Rajan, K., 2006. *Biocapital: the constitution of postgenomic life*. Durham: Duke University Press.

Swidler, A., 1986. 'Culture in action: symbols and strategies', *American Sociological Review* 51 (2), pp. 273–86.

Terranova, T., 2000. 'Free labor: producing culture for the digital economy', *Social Text* 18 (2), pp. 33–58.

Thorpe, C., 2008. 'Political theory in science and technology studies', in Hackett, E., et al., (eds) *The handbook of science and technology studies*. Cambridge: MIT Press, pp. 63–82.

Tocchetti, S., 2012. 'DIYbiologists as "makers" of personal biologies: how MAKE Magazine and Maker faires contribute in constituting biology as a personal technology', *Journal of Peer Production* 1 (2). http://peerproduction.net/issues/issue-2/peer-reviewed-papers/diybiologists-as-makers/, accessed October 2012.

Turner, F., 2006. *From counterculture to cyberculture*. Chicago: University of Chicago Press.

Turner, F., 2009. 'Burning Man at Google: a cultural infrastructure for new media production', *New Media and Society* 11 (1–2), pp. 73–94.

US Supreme Court, 1980. *Diamond v. Chakrabarty*, 447 U.S. 303.

Venter, C., 2005. 'Craig Venter: a voyage of DNA, genes and the sea', talk given at the TED Conference. http://www.ted.com/talks/craig_venter_on_dna_and_the_sea.html, accessed December 2008.

Venter, C., 2007. *A life decoded: my genome, my life*. New York: Viking.

Vettel, E., 2008. *Biotech: the countercultural origins of an industry*. Philadelphia: University of Pennsylvania Press.

Vise, D. and Malseed, M., 2006. *Google story*. New York: Delta.

Waldby, C., 2000. *The visible human project: informatic bodies and posthuman medicine*. New York: Routledge.

Ward, M., 2010. 'Tech now: life hacking with 3D printing and DIY DNA kits', *BBC News*. http://news.bbc.co.uk/2/hi/technology/8595734.stm, accessed October 2012.

Weber, M., 2003 (1905), *The Protestant ethic and the spirit of capitalism*. New York: Courier Dover.

Willinsky, J., 2005. 'The unacknowledged convergence of open source, open access, and open science', *First Monday* 10 (8). http://firstmonday.org/htbin/cgiwrap/bin/ojs/index.php/fm/article/view/1265, accessed October 2012.

Wohlsen, M., 2008. 'Do it yourself DNA: amateurs trying genetic engineering at home', *The Huffington Post*. http://www.huffingtonpost.com/2008/12/25/do-it-yourself-dna-amateu_n_153489.html, accessed June 2010.

Wohlsen, M., 2011. *Biopunk: DIY scientists hack the software of life*. New York: Current.

Wolinsky, H., 2009. 'Kitchen biology', *EMBO Reports* 10 (7), pp. 683–5.

Zilsel, E., 1945. 'The genesis of the concept of scientific progress', *Journal of the History of Ideas* 6, pp. 325–49.

Ziman, J., 2000. *Real science: what it is, and what it means*. Cambridge: Cambridge University Press.

Index